BLACK HOLES
QUASARS
AND OTHER
MYSTERIES OF THE
UNIVERSE

Dedication

To Beverly

Also by the Author from TAB BOOKS Inc.

No. 1505 *Understanding Einstein's Theories of Relativity: Man's New Perspective on the Cosmos*

BLACK HOLES
QUASARS
AND OTHER
MYSTERIES OF THE
UNIVERSE

BY STAN GIBILISO

TAB BOOKS Inc.
BLUE RIDGE SUMMIT, PA. 17214

FIRST EDITION

FIRST PRINTING

Copyright © 1984 by TAB BOOKS Inc.

Printed in the United States of America

Library of Congress Cataloging in Publication Data

Gibilisco, Stan.
Black holes, quasars, and other mysteries of the
universe.

Includes index.
1. Astronomy—Popular works. I. Title.
QB44.2.G53   1984   523   82-19298
ISBN 0-8306-1525-3 (pbk.)

Cover photos courtesy of Celestron International, Torrance, CA, U.S.A.

Contents

Acknowledgments

I WISH TO THANK THE FOLLOWING PEOPLE AND institutions for their help in obtaining photographs for this book: Dr. Mark Littman and the staff of the Hansen Planetarium; Dr. Paul Routly of the United States Naval Observatory; Dan Brocious and the Smithsonian Institution, Fred Lawrence Whipple Observatory; the National Aeronautics and Space Administration; the California Institute of Technology, Hale Observatories, Mt. Palomar and Big Bear Solar Observatories; and Mt. Wilson and Las Campanas Observatories, Carnegie Institute of Washington.

Introduction

SINCE NATURE FIRST DEVELOPED THE CAPACITY for reasoning and intelligent thinking in our species, we have been fascinated by the nighttime sky. On this, the third planet in orbit around a quite ordinary star, we can often see through the atmosphere into the heavens. We have always felt a certain sense of awe at the vista above us on a clear night. Long ago, people had little or no idea what was up there, but their feelings about it must have been similar to our own.

Within the past few centuries, our knowledge about the cosmos has multiplied. We have built powerful magnifying devices that make objects in the sky look thousands of times nearer and brighter. Photographic techniques have increased the ability of our telescopes to see the most faint wisps of stars being born, galaxies in formation, and some things about which we know almost nothing. We can now look at the cosmos at wavelengths we cannot see; with space rockets we can observe the sights which, until recently, our planet's air forbade us to see. But the strangeness of the universe does not seem to diminish as our knowledge grows wider.

Instead, the mysteries get deeper and more complex.

The distances to the stars are too great for us to consciously comprehend. At the other physical extreme, our own bodies are made from particles too tiny for us to envision. Interactions among these particles enable our minds to ask, "What is our place in the midst of all this?"

Astronomers and cosmologists of today believe that the universe originated in a great explosion, more violent than any hydrogen bomb, and brighter than any sun. The ultimate fate of the universe is not known. But we believe nature has its plan, and if we are patient, we can come closer to an understanding of it. In Chapter 1, we will explore the ancient as well as the modern theories of the structure and evolution of the cosmos.

Approximately 4 or 5 billion years ago, a cloud of dust and hydrogen gas, spinning in a vast whirlpool, collapsed under the force of its own gravity. From this cloud, the sun and planets were formed. Similar events have taken place over 100 billion times before in our galaxy alone. And there

The comet Ikeya-Seki, just before sunrise. Comets are thought by some scientists to have scattered the debris of rudimentary life throughout the universe. (U.S. Naval Observatory photograph.)

are billions of galaxies in the universe. How did the galaxies form in the manner they did? What will become of the stars? In Chapter 2, we will see what scientists have found in their search for answers to these questions.

In the 1960s, strange, unbelievably bright and distant objects were found in the heavens. Their energy output is greater than that of the galaxies. They are moving away from us at tremendous speeds, in some cases many thousands of miles per second. They are known as quasi-stellar radio sources, or quasars. Also during the 1960s, pulsing stars were discovered. They at first seemed so bizarre that astronomers thought they were the signals of intelligent beings. These are the pulsars. The radio telescope was instrumental in the discovery of these objects; we might never have noticed them had we been content to explore space with visual telescopes alone. In Chapter 3, we will examine the variety of apparatus that scientists use to view the cosmos. We will look at the contemporary explanations for the phenomena of the quasars and pulsars.

All fields of scientific endeavor have something in common with other pursuits—"fuzzy" areas of method and knowledge—and the black hole most certainly occupies an intuitive zone that is partly pure mathematics, partly observational astronomy, partly experimental physics, and perhaps even a little bit of abstract philosophy. In Chapter 4, we probe the anomaly of the space-time singularity, where matter, energy, space, and time disappear.

Perhaps the deepest reason why we pursue the secrets of the universe originates with a fundamental curiosity about ourselves. Where did we come from, and where are we going? If we reach the stars, what will we find? Is there another intelligent civilization somewhere? This is, and always has been, a topic for science fiction. But what are the facts? In Chapter 5, we look at the greatest mystery of all: the matter from which our bodies are made, the processes by which matter evolved to create us, and the likelihood that the miracle of life has occurred on other planets. We will explore the methods by which we might contact such other beings.

The ability to reason carries an unquenchable desire for knowledge. Questions must be answered. Answers are being found concerning the nature of our universe. Anyone can, and should, know what our astronomers, astrophysicists, and cosmologists are doing. This book is not a theoretical work, but a message. Almost anyone can follow and understand this message. But it is a message containing more questions than answers. That, however, befits our inquisitive nature: It is the questions that make us think!

Chapter 1

Looking at The Universe

WHEN YOU GAZE AT THE SKY ON A CLEAR, moonless night, far from city lights and the noises of civilization, you get the feeling that the universe must be a big place! Ever since the first ancient men, women, and children saw that sky, people have been trying to figure out just what they were seeing—its anatomy, its evolution, and its ultimate fate. Huge observatories and radio telescopes have been built in the effort to probe into the secret diary of our universe. Some scientists have devoted their whole lives to the pursuit of this knowledge. And no doubt there will be more observatories, more radio telescopes, more theoreticians and experimentalists, who put their minds and instruments to the task.

Why all this fascination with something so removed from the practical realities of life on this world? Answer that for yourself as you lie on a hilltop and look at the distant stars on a clear, moonless night.

WHAT IS THE UNIVERSE?

The term universe originally meant "all that is." It still carries that meaning, but in a somewhat incomplete sense—there may be more! We can speak of our universe, but perhaps there are other universes, too, existing in different dimensions of time, and operating in ways that might be totally unreal to us. The study of the structure of "all that is" is called cosmology; the most far-flung theoretical branches of cosmology involve models of alien universes. The study of the birth and evolution of the universe is called cosmogony.

According to the most widely accepted theory of today, our universe was created in an incomprehensibly brief and brilliant moment, several billion years ago. A cosmic explosion took place, and all the matter and energy in the universe was thrust outward from a single point in an otherwise void space. As the matter expanded and cooled from its initial searing temperature of trillions of degrees, the stars and galaxies eventually formed. Some of the stars had planets orbiting them. In one quite ordinary-looking galaxy, a spherical chunk of rock and metal about 8,000 miles in diameter orbited a yellowish, rather small star. Creatures

1

came to populate the surface of this planet. One species gave their home a name: Earth.

From this vantage point, a roughly spherical piece of matter orbiting a small star about three-quarters of the way from the center to the edge of a pinwheel-shaped congregation of 200 billion stars, intelligent creatures seriously try to unravel all the mysteries of the universe. It is believed that the universe has no boundaries, but is probably finite in extent, just like the surface of our own planet. Finite, yes, but vast—billions of light years across.

What is the universe? After all the discoveries have been made (if they ever are), we'll probably be able to say, with complete confidence, only that the universe is the surreal place in which we live.

THE EARLIEST MODELS OF THE UNIVERSE

When the sky is clear and there is no moon to hinder the visibility of the stars, you get a feeling of great depth as you look at the heavens. Early stargazers might have seen this depth, and decided that space was a huge expanse. But most ancient philosophers believed that all the stars were attached to a great sphere around the central planet, the earth. Some thought that the sphere was opaque, with little holes through which some outside light shone. Some thought that the stars were little fires on the inside of the sphere. No one asked, at least out loud, what might lie beyond the sphere. The sun was believed to hang on this sphere, too.

Eventually, observers noticed that there were some peculiarities about this model of the universe. The sun did not always stay in the same position with respect to the stars. Some of the stars, moreover, seemed to move among the stationary majority. These moving stars were called planets, or wanderers. They were assigned separate spheres of their own. The moon had its sphere, and the sun its sphere. Figure 1-1 shows one of the earliest cosmological models. The earth was assigned the center of all creation—there, after all, dwelt man!

As all of these spheres turned at their different speeds in the heavens, some ancient philosophers believed that they must create noise as they rubbed together. This noise would, of course, be very faint, since it originated so far away; only a trained ear would be able to hear the sound. Some said they could hear it, and that it sounded like beautiful music. This could only be the sound of the angels of the other members of the heavenly host, orchestrating the harmonious operation of the universe!

The Universe of Ptolemy

The model shown in Fig. 1-1 was soon seen to be an oversimplification of actual matters. According to the old model, all the planets would maintain a constant and uniform motion, always in the same direction, with respect to the background of stars. However, observers noted that the planets did not obey this law. Instead, they occasionally did a reverse. Some planets reversed their paths momentarily, more often than others. Plotting the position of a planet with respect to the stars, over a period of days or weeks, revealed this loop, as shown in Fig. 1-2. An astronomer of ancient times, Ptolemy, developed a model to explain this retrograde motion. This theory, called the Ptolemaic model of the universe, endured for centuries.

Rather than following a perfectly circular orbit around the earth, the planet was assigned an orbit consisting of two independent motions. These were called the deferent and the epicycle, and are shown in Fig. 1-3. The deferent followed a circular path around the earth. The epicycle, containing the actual path of the planet, revolved around the deferent. No reason for this was given; the model was simply invented to fit the observed facts. When the deferent and epicycle sizes were given the proper ratio, the theoretical model proposed by Ptolemy was an excellent approximation of the actual situation as seen from the earth, the center of the universe. But there were still small errors in some cases. Even in ancient times, skeptics could be heard as long as they did not deviate too much from accepted dogma! To compensate for these errors, additional epicycles were added, within the original ones. Thus there could be second-order epicycles, third-order epicycles, and so on. How complex were the motions of the planets! The network of sub-epicycles became almost overwhelming, and it

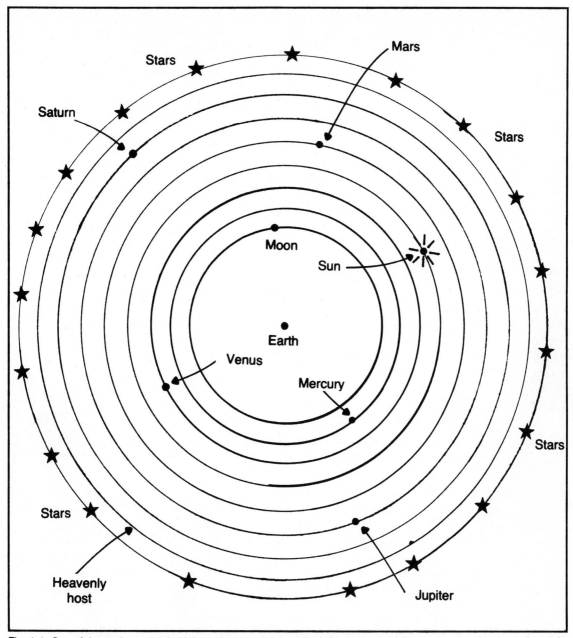

Fig. 1-1. One of the earliest models of the universe put the earth at the center, and the moon, sun, planets, and stars in concentric orbits around our planet. A place was even provided for the heavenly spirits.

provoked cynicism among the ancient philosophers. One king said that he might have given some advice at the time of creation, if he had had the opportunity! It seemed odd that the gods would be so messy, and that such a beautiful universe could be so horribly eccentric.

The geocentric theory, meaning "earth-centered," was made to fit the observed facts, without

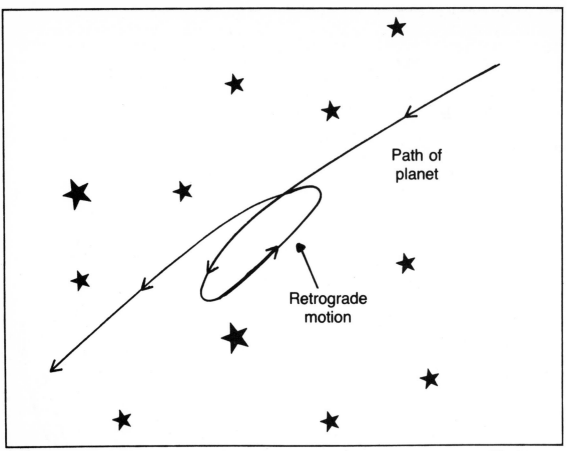

Fig. 1-2. The actual path of a planet, with respect to the stars, has an occasional loop. The universe model of Fig. 1-1 cannot account for this.

the realization that perhaps the whole basis for cosmology ought to be rehashed. It was perhaps too much for man to swallow: How could he be so mistaken? In ancient times, theologians dictated the beliefs of astronomers, and it seemed impossible that God could mislead them. If not the earth, what could be at the center of the universe?

Even today, we see cases in which emotionalism overrides rationality in scientific thinking. The ancients preferred to complicate matters without limit, rather than to admit that they might be fundamentally wrong.

Copernicus, Tycho, and Kepler

In the sixteenth century, Nicholas Copernicus published his theory of the universe. In this model,

the earth was not at the center of all creation, but was described as one of seven planets—Mercury, Venus, Earth, Mars, Jupiter, Saturn, and Uranus—revolving around the sun. Other astronomers had suspected, or perhaps wished, that this was the case, but none had dared to stand up and voice his belief in public. The penalty was severe. Some had been branded as heretics.

Copernicus was unable to prove his theory conclusively enough to convince the Establishment. How could the earth be in motion? What force would push it around the sun? Would this force not have great effects on the surface of our planet? How ridiculous, the Church thought, to believe such nonsense!

Copernicus, however, reasoned that the other

planets must revolve for some reason, whether about the earth or the sun; and no god could be seen pushing them. Why should the earth be so special?

The motions of the planets were recorded during the sixteenth century by an astronomer named Tycho Brahe. In his later years, his assistant was a man named Kepler. Kepler formulated a theory that became known as Kepler's Laws. They were published early in the seventeenth century.

Kepler deduced that the planets must, in fact, orbit around the sun, and not around the earth. But the orbits did not coincide with perfect circles. Instead, they were ellipses, with the sun at one focus. Kepler observed that the planets move faster when they are closer to the sun (near perihelion) and slower when they are farther from the sun (near aphelion). Also, the farther a planet was from the sun, Kepler saw, the longer it took to complete an

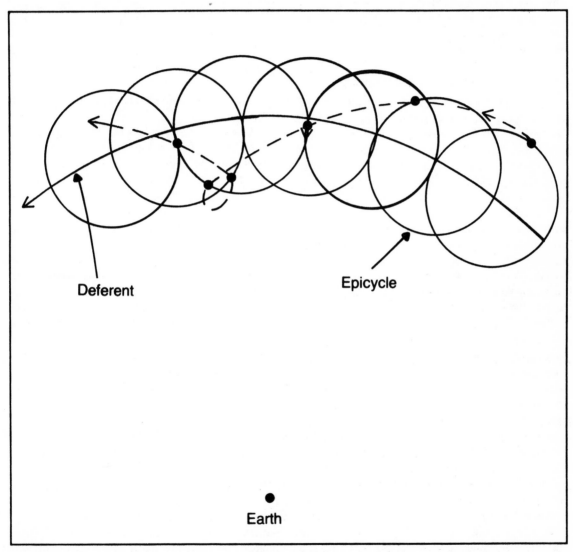

Deferent

Epicycle

Earth

Fig. 1-3. The ancient astronomer and philosopher Ptolemy explained the retrograde motion of the planets by postulating that their orbits around the earth actually consisted of two components: the deferent, or main orbit, and the epicycle, or secondary orbit.

orbit. There was a mathematical relation between distance and orbital period.

Figure 1-4 shows the laws of Kepler. Many theologians dismissed all this as heresy, and kept on believing the cosmology of Ptolemy. It had, after all, been accepted for more than a thousand years. Upstarts such as Copernicus, Tycho, and Kepler were not to be taken seriously, except for the fact that they presented a threat to the Establishment! This was true even though the new theory was less complicated than the older one, and explained the retrograde loops in the celestial paths of the planets with greater accuracy.

Galileo Galilei and Isaac Newton

An Italian astronomer named Galileo, who lived at the same time as Kepler, is credited with having built the first telescopes. Galileo also believed that the earth revolved around the sun, along with the other planets. Using his telescope invention, Galileo discovered mountains on the moon, satellites orbiting Jupiter, and rings around Saturn. He showed that objects having different masses fall with the same velocity when dropped. This earned him much scorn, even though he was able to demonstrate the fact in public. The Church took Galileo very seriously. He represented a grave threat to their theories, especially since he was able to offer proof of his views! Ultimately, Galileo Galilei was put under house arrest.

The heliocentric theory, or the idea that the planets orbit around the sun instead of around the earth, finally got general acceptance after the work of Isaac Newton. Newton, an English astronomer and physicist, provided the first model of gravitation, that strange, invisible force that causes all objects in the universe to attract one another. It is sometimes said that Newton got his idea while sitting in an apple orchard and watching an apple fall to the ground. (Some stories even say that an apple hit him on the head!) Whatever the actual impetus, Newton saw that the earth exerts a constant attraction on everything in its vicinity. All objects in the universe exert this same attractive force on the earth and on one another. Newton condensed the work of his predecessors and the results of his own research into his laws of motion, published in *Principia* in the seventeenth century.

Newton's theory of gravitation put forth the following three laws. First, any object at rest will stay at rest unless some force acts upon it. Any object with uniform motion will continue that motion unless an outside force intervenes. Second, a force on any mass produces an acceleration in the same direction as the force, in direct proportion to the force, and in inverse proportion to the mass of the object. Third, any action results in an equal and opposite reaction.

Newton's laws are still the basis of classical mechanics. Except when extreme forces, masses, or velocities are involved, Newton's laws are satisfactory to explain observed phenomena in our universe.

To explain the acceleration of falling objects, and the orbits of the moon, planets, and all other objects, Newton proposed that every object attracts every other. Given two objects X and Y, the force of attraction is proportional to the product of their masses $m_x m_y$, and inversely proportional to the square of the distance between them. This principle is shown in Fig. 1-5.

Finally it was evident why the planets could orbit the sun. They are all actually accelerating toward the sun because of gravitation, but this inward force is balanced by the inertia of each planet. Every object in orbit about the sun is in a state of free fall. The same is true of objects orbiting the earth, or objects in mutual orbit. The moon is falling around the earth. Jupiter's moons are falling around that planet. The existence of gravity became clear. Its actual mode of operation remained a mystery.

Orbits

Johannes Kepler showed that the planets do not orbit the sun in perfect circles. In fact, such a perfect orbit is impossible because of the mutual effects of gravitational fields among the planets. There is always some deviation from perfection. The moon's distance from the earth varies. The distance between the earth and the sun varies. Some asteroids have extremely elongated orbits. And some celestial objects are pulled into the solar

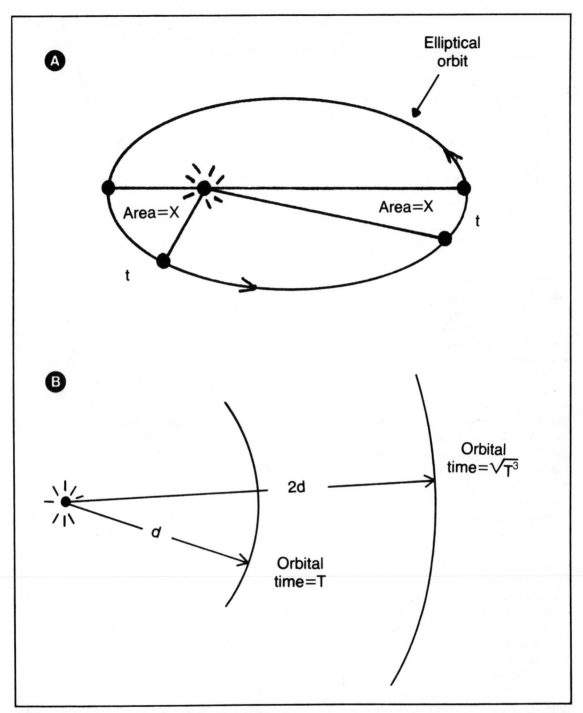

Fig. 1-4. Kepler found that the planets circumnavigate the sun in elliptical orbits. The area swept out by a planet in a given length of time, as shown at A, is always the same, no matter what part of the orbit it travels. The period of an orbit, T, increases as the 3/2 power of the distance, d, as shown at B.

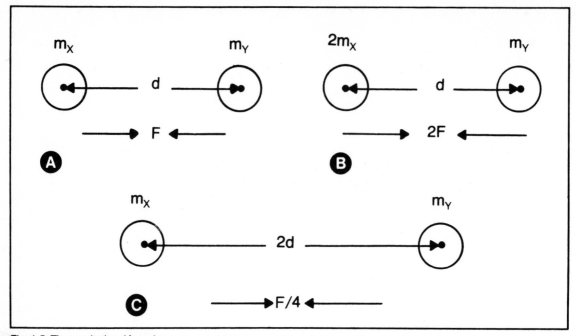

Fig. 1-5. The gravitational force between two objects is proportional to the product of their masses, and inversely proportional to the square of the distance between them. At A, two masses m_x and m_y, separated by distance d, have a gravitational force F. At B, if one mass is doubled, the force is doubled. At C, if the distance is doubled, the force is cut to one-quarter its previous value.

system by the gravity of the sun, make one near pass, and then depart never to return.

The path of an object from outside the solar system is not an ellipse. Instead, it is a parabola or hyperbola. Any object orbits any other in a path called a conic section. Figure 1-6 illustrates the four kinds of conic sections. These are the circle, ellipse, parabola, and hyperbola.

You can easily generate these conic sections using a bright flashlight on a large, smooth surface in a dark room or on a dark night. Standing on the flat surface of a parking lot, tennis court, or gymnasium, shine the flashlight straight downward. The perimeter of the dim, wide portion of the beam (not the bright central part) forms a circle. If you tilt the flashlight slightly, the circle becomes elongated. This is an ellipse. The more you tilt the flashlight, the greater the eccentricity of the ellipse. At a certain tilt angle, the far end of the ellipse will disappear, no longer reaching the ground. The edge of the wide-angle area is then a

parabola. If you tilt the flashlight so that it is shining horizontally, the edge of the wide part of its beam lands on the ground in the shape of a hyperbola.

If an object happened to have a perfectly circular orbit, Newton realized that a slight nudge would send the object into an elliptical orbit. If the nudge were stronger, the ellipse would be very elongated. If the push were sufficiently forceful, the object would achieve escape velocity and leave the gravitational influence of the central mass; its orbit would describe a parabola. A push of great force, much more than the minimum needed to produce escape velocity, would send the object flying away along the curve of a hyperbola. All of these orbital configurations occur in space, with (as we have said) the exception of a perfect circle.

The central mass is not unaffected by an object in orbit. Even the tiniest satellite, orbiting around the earth, causes the planet to move slightly. Two equally massive objects would, if placed in proximity, orbit each other around a common center, mid-

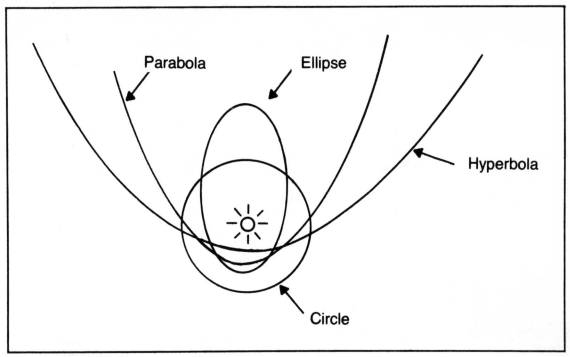

Fig. 1-6. Objects may orbit the sun in a circle, an ellipse, a parabola, or a hyperbola.

way between the objects. This is shown in Fig. 1-7A. When two mutually orbiting masses are unequal in size, the center of their orbit is closer to the larger mass (Fig. 1-7B). Although Sputnik, Project Mercury, and communications satellites do not disturb the earth very much, the effect does occur. The moon causes the earth to be perturbed to the extent that the orbital center lies 2,900 miles from the actual center of our planet.

Absolute Space and Time

Newton developed his theory of gravitation to explain the motions of celestial objects. He reasoned that everything in the universe was influenced by the force called gravitation. Even the most distant stars must obey Newton's principles. What implications about the universe could be derived from this theory? If every object in the universe exerts a force on every other, certainly this might have significance in the structure and evolution of the cosmos.

Newton apparently did not see the link be-

tween his laws and the existence of inertia. Why does a mass remain at rest unless acted upon by an external force? Newton decided that the reason lay in the absoluteness of space itself. A great fixed emptiness, he thought, was responsible. This is where Newton departed from his empirical ways, and began to act as a theoretician and philosopher. His conclusions began to be based on speculation more than observation. That is always dangerous territory for a scientist! Why does the moon not fall down to the earth? Newton believed that the moon was revolving around the earth, with respect to the viewpoint of absolute space. The same was true of the earth with respect to the sun, and with all the other planets as well. The idea of absoluteness—a last word for the determination of motion—in space was lent support by the Foucault pendulum experiment.

This experiment was constructed to prove that the earth rotates on its axis. It can still be performed, as shown in Fig. 1-8. A heavy pendulum with a long wire support must be suspended from a

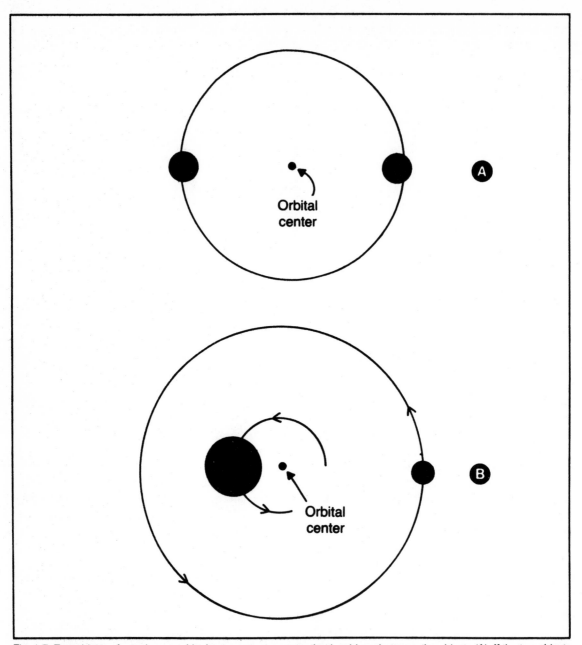

Fig. 1-7. Two objects of equal mass orbit about a common center that is midway between the objects (A). If the two objects have different masses, the orbital center is closer to the object with the greater mass (B).

room with no air currents to disturb the swing of the weight. As the earth rotates, the plane of the pendulum swing also rotates. At the pole, the plane of the swing will be seen to go once around in 24 hours. At the equator, no change will be seen. At some intermediate latitude, the plane of the swing will turn at an intermediate rate of speed. The only explanation, at the time this experiment was first done, seemed to be that space had absolute properties, and that the pendulum was influenced by these properties.

Having satisfied himself that space was absolute, Newton reasoned that time must be the same way. According to Newton, time flowed smoothly, and was not affected by anything. This seemed logical enough. There was no way, at that time, to show otherwise! But this idea was not based on any observed facts, other than the apparent constancy of the functioning of a properly aligned clock. Relativistic physics shows that this concept is far from the truth: Time can flow at different speeds. Not only that, but space can be stretched and squeezed,

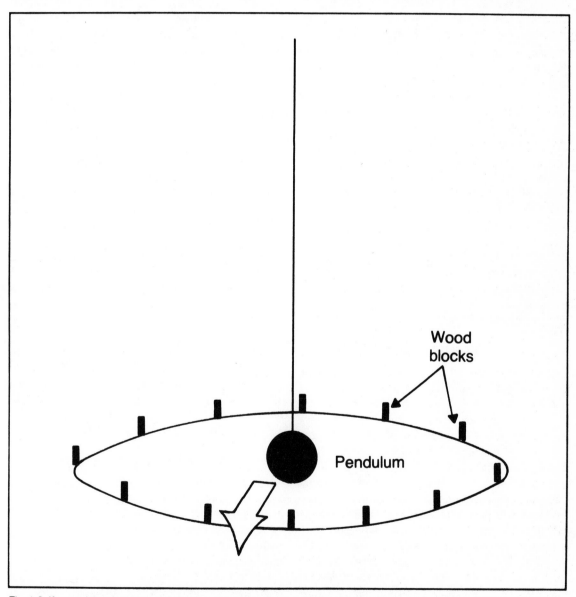

Fig. 1-8. If a pendulum is swung anywhere on the earth except at the equator, the plane of swing will slowly rotate. Eventually, all of the wooden blocks in this illustration will be knocked over by the swinging pendulum.

and mass can change! But this was not discovered until the twentieth century.

In the generations immediately following the acceptance of Newton's laws of gravitation and motion, scientists tried unsuccessfully to find an exception. Methods of measurement became more and more refined. Inquisitiveness was encouraged, and not frowned upon or suppressed. Finally, one nagging exception was found to Newton's principles, one fact that did not fit the theory. Like a buzzing mosquito, it presence could almost—but not quite—be ignored. In the world of scientific logic, one fault is as good as a thousand.

The perihelion, or point of closest approach to the sun, in the orbit of Mercury changes position by about 43 seconds of arc per century. (A second of arc is 1/3600 of a degree.) In 1845, an astronomer named Leverrier observed this precession in the perihelion of Mercury. There was no doubt: It was real. And there was also no doubt: Newton's laws could not explain the reason for it.

Attempts were made to explain this phenomenon, but no satisfactory model could be found until Einstein predicted it with his theory of relativity. For a time, scientists got on with the rest of Newton's theories, and their implications, hoping that eventually a solution would be found, that the mosquito would be swatted.

HOW BIG IS BIG?

Before we continue to look at theories of the universe, let us examine the sorts of distances involved in any discussion of the cosmos. We can speak of a light year almost as if such a distance were a few inches. Our universe is, according to the most recent models, billions of light years across. But what is a light year? And how big is a billion?

You have probably heard that light travels very fast, about 186,282 miles per second. In a minute, a beam of light gets 11,177,000 miles, or about an eighth of the way from the sun to the earth. In an hour, light can travel about 671 million miles. This is approximately the distance from the sun to a point midway between the orbits of Jupiter and Saturn. In a day, a ray of light travels about 16 billion miles; that's about twice the diameter of the known solar system. In a year, a ray of light travels about 5.8786 trillion miles. That distance is called a light year, for obvious reasons. It is an incomprehensible distance. This is why we toss it around with such indifference when we discuss cosmology; if we fully understood the magnitude of what we are saying, we might go utterly mad!

The nearest star, Proxima Centauri, is about 4.3 light years away. There are many stars within a radius of 100 light years of our sun. The whole Milky Way galaxy measures about 100,000 light years from edge to edge. With the most powerful telescopes, we can see objects (actually, they must be photographed) that are billions of light years from us.

What is a billion? Well, suppose that you had a cubical cardboard box measuring exactly one meter—about three feet—along each edge. Now imagine that you also had a huge pile of tiny blocks, perfect cubes measuring just a millimeter along each edge. (Look at the metric scale of a ruler and see how small a millimeter is.) Now imagine, if you can, the task of stacking these cubes in perfect alignment, one at a time, in the big box. When you finished the job, you would have stacked a billion little cubes: a thousand high, a thousand wide, and a thousand deep. How long would that take? At one cube per second (that's hurrying right along), the job would require nearly 32 years.

Try that ten times in a row: 320 years of stacking little cubes in big boxes. If each little cube represents one light year, then you would, after 320 years, begin to appreciate the distance *in light years* to some of the more remote galaxies and quasars. How far is that in miles? Never mind!

ALBERT EINSTEIN'S THEORIES

By the end of the nineteenth century, it seemed that physics, and hence astronomy, was on the edge of a new age of certainty. Only a few loose ends needed to be tied up, and the theory of physical reality would be complete. Total knowledge—the kind of knowledge that Ptolemy created—seemed almost at hand. Then, a few brilliant scientists blew it all away. Perhaps the most famous, although certainly not the only, one was Albert Einstein. His theory of relativity changed man's view of the cosmos, from the most fundamental behavior of matter and energy to the general concept of the universe.*

In his youth, Einstein was not regarded as a particularly brilliant individual. He seemed unable to apply himself to the practical matters of an ordinary world. Instead, he preferred to daydream with his favorite tools, the mind, paper, and pencil. Einstein had a special ability with mathematics, and it is said that he was also quite talented with the ladies! The overwhelming drive in the man's life, however—the thing that demanded the most of his effort—was his unquenchable curiosity about the workings of the cosmic machine.

The main axiom of the theory of relativity is that the speed of light in free space is always the same, regardless of the point of view of the observer, as long as the observer is not accelerating. This single axiom led to all the conclusions and ramifications of relativity theory, including time dilation, spatial distortion, and mass distortion. No point of view has any special properties over any other, as long as that point is not accelerating. According to Einstein, there was no such thing as absolute space. The only constant, he said, was the speed of light.

Isaac Newton's idea of absolute space, ironically, contained the seeds of its own destruction. The nagging belief that space must have some property of being "at rest" led to the proposition that electromagnetic waves, such as light rays, were propagated via some medium. The wave theory of light had been recently discovered around the end of the nineteenth century; scientists reasoned that the wave oscillation must be carried in a sort of cosmic jelly. How could waves travel through empty space? Scientists reasoned that light must be propagated like sound, by some kind of vibration of an identifiable substance. This medium, which was assumed to permeate all the universe, was called the ether. A great search was carried out to find the absolute velocity of the ether

relative to the earth. If this speed and direction could be found, then the ultimate velocity of the earth in the universe could be determined.

How would experimenters go about finding this velocity? It was reasoned that the speed of light should be constant in the ether, and that if the earth were moving through the cosmic jelly, we should notice that the speed of light is greater in some directions than in others. A beam of light should, it was reasoned, appear to travel faster when sent in a direction opposite to the motion of the earth. Light should seem, it was surmised, to move more slowly when sent in a direction nearly the same as that of the earth through the ether. But all efforts to find any variation in the speed of light met with failure. No matter what the direction, the speed of light was always the same. Could it be that the earth was actually at rest in the ether, and that the earth was the center of the universe? Scientists thought not.

Puzzled physicists at first tried to explain this result by arguing that the earth, and in fact all stars and planets, must pull some of the ether along. This would stand to reason. It would occur only in the vicinity of the earth. But eddy currents would also naturally take place in such a situation, and this would affect the speed of light at least once in a while. Einstein looked at the ether theory and dismissed it as too complex and unfounded. He saw no reason for the ether to exist at all. To him, space was not absolute; it was just what it seemed to be: empty!

Albert Einstein published his special theory of relativity early in the twentieth century. He was working at the Swiss patent office at the time. A few years later, he published his general theory of relativity, which dealt more directly with the structure and evolution of the universe. Einstein believed that the universe was finite, and yet unbounded. This strange idea has become more and more commonly accepted since he first proposed it in a paper in 1917.

The thrust of the general theory of relativity is

---

* A thorough discussion of relativity for the layman is given in *Understanding Einstein's Theories of Relativity: Man's New Perspective on the Cosmos* (TAB Books, Inc., 1983).

that the presence of a gravitational acceleration causes space to be non-Euclidean. All the laws of Euclidean geometry are thus invalid. Near a celestial object, such as the earth, the path taken by a ray of light is not straight, but curved. In the overall sense, the totality of masses in the universe causes space to be folded back on itself.

NON-EUCLIDEAN SPACE

Many physicists and engineers were incredulous at the idea that the principles of Euclid might not hold in the universe. Students are taught Euclidean geometry to this day! Two parallel lines must never, thought the scientists, intersect! But in non-Euclidean space, parallel lines cannot exist. Flat planes or straight lines cannot exist either.

We might get some idea of the curvature of space by imagining the surface of our own planet. Ancient people thought that the earth was flat. Sailing on the open sea, it seemed obvious that the earth must be flat. But there were subtle indicators that showed otherwise. A ship, sighted from a distance, seemed mysteriously to ride low in the water. The sun shone from different angles at different latitudes. Magellan set out to prove that the earth was spherical; he proposed to sail around the globe. Although Magellan himself did not survive the journey, a few of his comrades did. No doubt they believed.

For a three-dimensional spatial continuum to have curvature, there must exist a fourth dimension. The mental hurdle of imaging a fourth dimension is difficult, for we cannot imagine a space of more than three dimensions. In a four-space, it would be possible to orient four rods in a mutually perpendicular arrangement. Four coordinates would be needed to uniquely define a point. From the vantage point of a four-dimensional being, a three-dimensional space would seem infinitely thin and confining. But three-space could be curved! Just as the two-dimensional surface of a balloon is curved in three dimensions, Einstein reasoned, so is the three-dimensional "surface" of our universe curved with respect to four-space.

In a "spherical" universe of this kind, if a space traveler were to set out in one direction and travel far enough, he would eventually return to the solar system from the opposite direction! According to Einstein's calculations, the universe had a diameter of approximately 100 million light years. This estimate was based on the known masses and distribution of the stars. At that time, it was not yet known that the stars were arranged not uniformly, but in great clusters called galaxies, separated by thousands or millions of light years. Today, the diameter of the universe is estimated at a much larger value—on the order of 15 to 30 billion light years. No cosmic Magellan will circumnavigate our universe and return with his story to the earth! Our planet would long since have been destroyed by the evolution and death of the sun!

THE TOTALITY OF MASSES

We have seen that Isaac Newton believed in absolute space and absolute time. Modern scientists know better. Rapid velocities cause time to progress more slowly, spatial distances to become shortened, and masses to increase. Extreme acceleration can actually curve the geometry of space significantly. But how do we explain the existence of inertia? How can we explain the results of the Focault-pendulum experiment?

Imagine, for a moment, that you were the only object in the entire universe. You would of course be dressed in a space suit with a good supply of oxygen. This might become reality if you could somehow rip yourself away from the three-dimensional continuum of our cosmos, and build your own little closed universe. You would be weightless, since there would be no other masses to put any gravitational pull on you. You would see no stars, and no galaxies—nothing at all! The emotional effects of such a situation might well render you quite irrational, but suppose you were a particularly strong person in this respect. Imagine yourself pulling out your little rocket gun—the device you use to move around during space walks—from one of the many pockets in your outfit. Would you begin to move when you fired the gun? Would you feel the inertial effects of the thrust? Would an accelerometer show it?

This question is, of course, difficult to answer

from experience! Nobody has ever carried out such an experiment. (And if they could, they would never be able to return and tell us the results!) But we can argue, in a theoretical sense, that there would be *no* inertial effects. In a universe containing only one object, there would be no basis for determining motion. Motion is relative. Your situation before, during, and after the firing of the rocket gun would be exactly the same. First, there would be you, floating in the void. Then there would be you, floating in the void, with a bright flame spurting from the nozzle of a rocket gun. Finally, there would again be you, floating in the void. There would be no change in your motion, because there would be nothing with which to ascertain motion. So there would be no inertia. Newton, imagining this experiment, would probably say that you would get the feeling of inertia, because you had changed your motion with respect to empty space! But how can you do that?

The only logical, or at least intuitively acceptable, conclusion is that you would feel no inertial change. Your motion would remain the same, since there would be no basis for determining it in the first place.

How does our universe differ from an empty one, containing only yourself? The answer is trivial: There are other things in the actual universe. A lot of other things! Thus, these other things—the earth, the moon, the sun, the planets, the distant stars, galaxies, black holes, and quasars—all of them, in their totality, are responsible for the fact that an object tends to maintain its motion unless acted upon by an external force. The totality of masses in the cosmos is responsible for the fact that you must push a motorboat to get it moving across a lake! The distant masses in the universe are the motivation for the moon to keep from falling into the earth because of mutual gravitation.

Einstein was not the only scientist to put forth this idea. But he stood out from the others in his field, because he pursued it, and got his idea of the universe in accordance with it. The totality of matter in the universe, by making inertia a reality, also makes mass itself a reality, since mass is measured and determined on the basis of inertia! Even the gravitational constant, which determines the amount of force produced by a gravitational field of a given intensity on a given mass, may depend on the mass and density of the universe. The mass of the universe, Einstein showed, determines its size, and this determines the rate at which time progresses.

One consequence of the theory of relativity is rather unpleasant for those who like to envision interstellar and intergalactic voyages. It is not possible, according to Einstein, for any material object (such as a space ship) to travel faster than the speed of light. No matter how powerful the engines of a space vehicle can ever be made, the speed of light represents an insurmountable barrier. Many stars are hundreds, or thousands, of light years away from us. The most distant galaxies are millions, or even billions, of light years distant. It would require many lifetimes to reach them. Although it may, someday, be possible to reach these far-off objects, a return to the earth following such as a journey would involve some psychological problems, as we shall soon see.

VERY FAR AGO

Ancient men probably never suspected, as they gazed at the sky, that they were looking at the past. When you look at the moon, for example, you are seeing what took place about 1.34 seconds ago. As you observe the sun (with a pinhole projection device, not directly!), you are seeing that star as it was about 8 minutes ago. The planets are light minutes or light hours away. The nearer stars are several lights years from your eyes.

As you lie on a hilltop and look up at a clear sky on a moonless night, you are actually looking into the distant past. One particular star may be, say, 10 light years away. Some photons of visible light energy left the fiery surface of that star when you were 10 years younger. These photons hurtled at 186,282 miles per second upward through the corona of the star, out from its disk, and into the surrounding darkness. After a few hours, the photons were well into the interstellar void. They survived the risk of collisions with particles of dust and asteroids. One yellowish star became nearer to these photons, and began to stand out from the other stars. A planet loomed. Clouds, blue sea, and green and brown land masses approached. On the dark side of the disk, a living being lay on a hilltop. The photons stabbed with incomprehensible speed into the eyes of this being, and resulted in impulses to its brain. The being realized: A star is visible.

As your eyes become adjusted to the dim light, you see a faint, hazy band that seems to stretch across the canopy overhead. Is it a cloud? No, it is a host of millions and millions of distant stars. With a pair of binoculars, many individual stars can be seen. Some of them are thousands of light years away. Many are obscured by interstellar dust clouds. (This is a fortunate thing; if not for these dust clouds, the center of the Milky Way galaxy would be so bright that we would have no nights for much of the year.)

A single, dim point of light comes to your attention. The photons that you see left this star while Napoleon's armies were battling the fierce winter on the road to Moscow. A fainter pinpoint is visible through the binoculars. The photons you now see left this star while the soldiers from inside the wooden horse were capturing the city of Troy! So legend tells us, at any rate.

A faint fuzzy ball is visible off to one side of the Milky Way. You must blink a couple of times to be sure it is really there. What is it? A cloud? No—it is a galaxy, entirely separate from ours. Through your binoculars, it has a somewhat oval-shaped appearance. It is called the Great Nebula in Andromeda. The photons from this galaxy left it two million years ago! This is before historical records were kept (Fig. 1-9). We see the past—now.

The speed of light, according to the theory of relativity, is the speed of time. It is the only absolute in the universe. Although there is some evidence that superluminal motions (speeds greater than the velocity of light) are possible under certain circumstances, we know that visible light, and all electromagnetic radiation, travels in space at the finite speed of 186,282 miles per second. Looking at the universe, we see the past. Light and other electromagnetic emissions provide the only win-

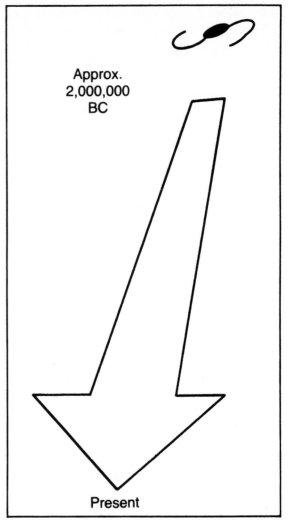

Fig. 1-9. Light arrives from the Great Nebula in Andromeda after having left that galaxy 2 million years ago.

dow to the cosmos for our telescopes and radio telescopes. Perhaps someday we will find a way to observe the universe as it is now, rather than as it was thousands, millions, or billions of years ago. But we are, in a sense, fortunate that light travels with finite speed. By probing the depths of the cosmos, we can get ideas about its evolution. We can look thousands, millions, and even billions of years into the past! This has been instrumental in the formulation of a model for the structure of our universe: how it began, how it is evolving, and how it will ultimately end.

SIMULTANEITY IS AN ILLUSION

The fact that the speed of light is the speed of time, as far as our own perception is concerned, has some strange implications. Time cannot be synchronized among different points in such a universe. On the moon, it is not the same time as it is on the earth, no matter how hard we try to make it so. Nor is it the same time on your nose as on your ear!

Suppose we planted an extremely accurate atomic clock on the moon, equipped with a broadcasting transmitter, just like the time stations you hear on a shortwave radio receiver. How would we, standing on the moon, set this clock? We would tune in an earth-based time station, such as the National Bureau of Standards station WWV in Colorado. We could then electronically synchronize the moon clock with WWV. Now, we would say with a sigh, we have found, or at least invented, a time standard for the moon. Our mission accomplished, we would return to the earth. What would we find when we checked the synchronization of the clocks from the vantage point of our home planet?

The clock readings would disagree! The earth-based clock, at WWV, would be ahead of the moon-based clock by 2.68 seconds. This would not be because of any malfunction of either clock, but the simple result of the spatial separation between the earth and the moon. Then, we might ask, what time is it, anyway? The radio signals from the moon-based clock can only travel at 186,282 miles per second, and the moon is 250,000 miles away from the earth. That is an error of 1.34 seconds. When we set the moon clock, the earth station was 250,000 miles away; that accounts for the other 1.34 seconds. The total discrepancy is thus 2.68 seconds.

The time depends, then, on the point of view. On the moon, we would probably use moon-based clocks. On the earth, we would of course use earth-based clocks.

It is fairly easy to see the trouble with putting our trust in the word "simultaneous" in space. When great distances are involved, it is impossible, obviously, to use this word. But what about small distances? Although the discrepancies are minute, they exist even within the room in which you now rest. They exist between your feet and your hands. A time discrepancy is present even among your brain and nerve cells; it is there between one side of an atom and the other; it is present even across such a tiny span as the diameter of a proton. Such time discrepancies are too small for us to perceive or, in some cases, even to measure. But they are always present! The word "simultaneous" means exactly synchronized. Exactly! That never happens.

How strange is the universe!

LONG-DISTANCE SPACE TRAVEL

When journeying to other stars and galaxies, problems will arise unlike any other that man has ever experienced. The nearest star to our solar system, Proxima Centauri, is more than four light years away. Suppose that a space ship could somehow be rapidly accelerated to 99 percent of the speed of light, or 184,419 miles per second. If this space vessel went to Proxima Centauri, 4.3 light years away, and immediately turned around and came back, it would arrive about eight years and eight months later. If the ship left in July of the year 2015, it would return in March, 2024. This is the situation as it would be perceived by earthlings.

But for those travelers bold enough to make the journey, a peculiar thing would happen: Time would be dilated by a factor of 1/7. According to the special theory of relativity, if v is the average velocity of a moving object and c is the speed of light in the same units as v, then time is dilated by a factor of $\sqrt{1 - v^2/c^2}$. If t_{earth} years pass on the earth, then t_{ship} years pass on the space ship, and

$$\sqrt{1-v^2/c^2}$$

$$t_{ship} = t_{earth} \quad \sqrt{1 - v^2/c^2}$$

This effect, called relativistic time dilation, occurs for clocks of all kinds, as well as all biological tissues, human brains, thoughts, and minds. Einstein predicted this effect by mathematical calculation, using only his axiom about the constancy of the speed of light from all points of view. This effect was experimentally demonstrated in the 1970s by scientists, using a jet airplane and two highly precise atomic clocks. In theory, time dilation occurs as you walk across a room or down the street! It is a tiny effect at such slow speeds, as you can calculate from the formula, but it does take place. At 99 percent the speed of light, time moves only 1/7 as fast as it does on earth. The space travelers would perceive only one year and three months during their journey. Their clocks would so indicate, and their bodies would so age.

This would be a strange experience, indeed! The difference between t_{earth} and t_{ship} over the journey would be seven years and five months! If you were 32 years old when you left, and your younger brother, age 29, stayed behind, he would be your older brother upon your return. You would be a bit over 33 years old, having aged a year and three months. He would be almost 38, having gotten eight years and eight months older.

As the velocity of a space ship approaches the speed of light, the term $\frac{v}{c}$ in the above equation approaches 1, and the time-dilation factor approaches zero. It is impossible to actually reach the speed of light, because the mass of the ship would increase to enormous values, making acceleration more and more difficult. But it might someday be possible to reach speeds very close to the speed of light. The time-dilation factor might be made very small—perhaps on the order of a millionth, or 10^{-6}. In such a case, intergalactic travel would become possible.

But wait a minute. No, wait a few million years! If a space ship took off for the Great Nebula in Andromeda, two million light years away from the solar system, it would not return to earth for a long while. Even at nearly the speed of light, the ship would not return by earth time until four million years had passed. If we took off for the Andromeda galaxy in a space ship with a time-dilation factor of 10^{-6}, we would be able to return to our home planet in only four years—to find what?

Imagine a space vehicle capable of accelerating and decelerating indefinitely at one gravity. This is about 32 feet per second per second. A long space voyage would not be a terribly uncomfortable experience aboard such a vessel. We could accelerate at one gravity until we got halfway to our destination; we could then turn our ship around 180 degrees, endure a brief period of zero acceleration, and decelerate the rest of the way at one gravity. Our bodies would be subjected to exactly the same force that they are used to on the earth.

An acceleration of 32 feet per second per second might not sound like very much. After one second, we would attain a speed of 32 feet per second, or about 22 miles an hour. After ten seconds, our speed would be 320 feet per second, or about 220 miles an hour. Jet airplanes accelerate almost this fast! But given a few months or years of acceleration at this rate, our speed would approach that of light. After one month, our speed would be 8.6 percent the speed of light. After six months, we would be moving at half the speed of light. As the speed of our vessel increased, the spatial acceleration necessary to produce one gravity would get smaller and smaller, although the force of the engines would remain the same. This would take place because of time, mass, and space distortion. Our speed would approach that of light, as shown in the graph (Fig. 1-10).

With this space vessel, a trip to another galaxy could be carried out within a human lifetime. The known universe could be circumnavigated in just a few decades. We could become cosmic Magellans, perhaps proving (at least to ourselves) that the universe is, indeed, shaped like a four-dimensional sphere with a three-dimensional surface. But upon our return to the solar system, what would we find? Would there still be any life on the earth? Would this life, if it existed, resemble human beings? After a complete circumnavigation of the universe, would the sun have completed its evolution, become a red giant, expended its energy, swallowed or incinerated our planet, and died? Probably! Then, future

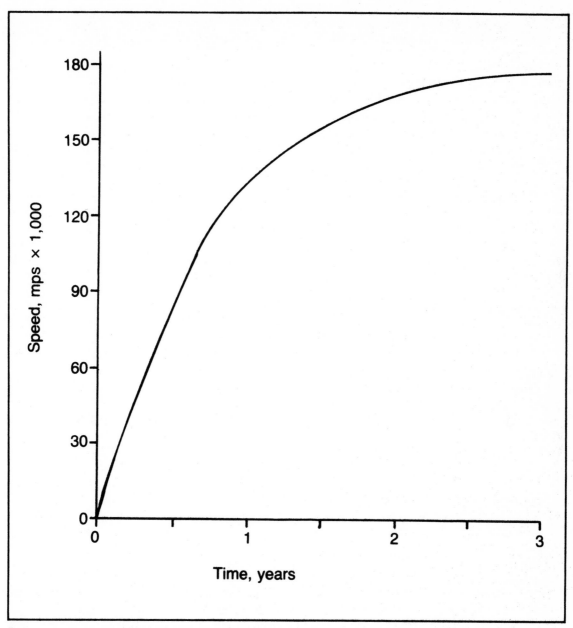

Fig. 1-10. If a hypothetical space ship were accelerated at 1 gravity, or about 32 feet per second every second, the speed would increase with time as shown here.

space travelers might ask, when confronted with such a journey, why go at all? The psychological implications are devastating. Such a trip would amount to time travel: a one-way ticket into the future, with no return flight ever available.

SIGNALS IN THE COSMOS

All electromagnetic energy, not just visible light, travels through space at the speed of 186,282 miles per second. Searches have been conducted for radio signals from the cosmos. Perhaps a message from another world is impinging upon our planet at this moment—and we might receive it, if we only knew in which direction we should point our radio telescopes, and at what frequency to listen! But any signals from another world must be signals from the past. The farther away the alien signals originate, the longer ago they were sent.

The discrepancy of time between the earth and a space vessel was evident even during the Apollo moon flights. From a distance of 250,000 miles, it took a few moments for the astronauts to reply to transmissions from the earth. If the astronauts had been orbiting Mars, the delay would have been several minutes, perhaps even as long as 40 minutes if the planet were on the far side of the sun. From the nearest star, the delay would be almost nine years. It is clear that, once space travelers are very far away, transmission to and from them will have to be of a one-way nature.

Actual communication via radio with beings on another world, orbiting another sun, is not practical for this reason. There might, by chance, be an intelligent civilization on a planet only 40 light years away from us, although this kind of fortunate coincidence is unlikely. A reply to a question to these beings would return after about one human lifetime. The distances among the stars proves, itself, to be a great obstacle to interaction among hypothetical civilizations in different star systems. But the time barrier compounds this problem.

Someday, perhaps a "subspace" or "hyperspace" mode of electromagnetic communications will be found. At present, this is the exclusive realm of science fiction. We might derive comfort, however, from the realization that fact is stranger than fiction. The more we learn about the universe, the less, in proportion to the whole, we know!

THE RED SHIFT

It was not always known that the strange, wispy pinwheels and other fuzzy lights in the sky were actually galaxies, just like our own, the Milky Way. Some scientists, as early as the beginning of the twentieth century, believed that the nebulae were small and relatively close by. In 1913, an astronomer named Vesto Melvin Slipher observed something peculiar about the light coming from some of these objects.

All stars have light with certain similar characteristics. When the light from any star is passed through a prism, so that the wavelengths are spread out into a rainbow type spectrum (Fig. 1-11), dark bands occur in certain places. The star is "black" at the wavelengths corresponding to these positions in the spectrum. This effect takes place because atoms in the gases of the star have very well-defined energy states, and this results in absorption of the light from the star at discrete wavelengths. The sun has a spectrum containing such absorption lines. Nearby stars, as well as the more distant ones, show these dark bands. While the pattern of bands varies a little from star to star, some of these absorption lines are always present. In every star, for example, there is a pair of lines corresponding to the element calcium. We can see and identify this pair of absorption lines in the spectrum of a large group of stars. We can even see it in the spectrum of a whole galaxy.

Vesto Melvin Slipher was not looking for, or expecting, what he found in 1913. But upon careful observation of the spectra of the galaxies, he saw that the characteristic pair of calcium-absorption lines was not quite in the right place. They appeared too far toward the red end of the spectrum. Their wavelengths were too long. Slipher did not immediately know what this meant, but he believed that it must be a significant result. This discovery was called the red shift. It was received with great interest and enthusiasm among astronomers and cosmologists.

What might cause this red shift? The most obvious explanation was that the red shift is the result of the Doppler effect, occurring with the light from the distant galaxies. When an object is rapidly moving toward an observer, the wavelength of an energy source on that object will seem to be shortened, as compared to the case when the object is not moving with respect to the observer. When an object is moving rapidly away, the wavelength of an energy source is lengthened as seen from a remote

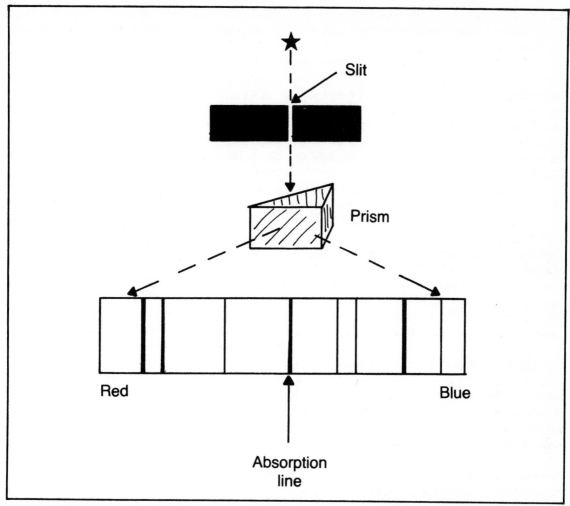

Fig. 1-11. The spectrum of a star contains dark lines, corresponding to light absorption caused by various elements in the atmosphere of the star.

observation point. This is called a red shift. It is the same effect that makes a train whistle sound higher in pitch as the train approaches, and lower in pitch as the train recedes.

Doppler shift results only from that component of motion toward or away from an observer. Sideways motion does not have this effect. If the red shift seen by Slipher was actually a Doppler shift, then all of the spiral nebulae were retreating from the solar system. Why would that happen?

The Doppler effect is not the only possible cause of a red shift in the spectrum of an object in space. A strong gravitational field can also produce a red shift. Perhaps some other effect, not yet discovered, is responsible for the red shift in the spectra of the spiral nebulae. But the most plausible explanation for most cosmologists, even to this day, is that the red shift is caused by radial motion, and that the spirals are moving away from us.

DISTANCES TO THE SPIRAL NEBULAE

In the early part of the twentieth century,

when Vesto Melvin Slipher announced his findings, there was still much disagreement on what the spiral nebulae actually were. Some thought that they were vast, enormous, distant congregations of individual stars. Others believed that they were interstellar gas and dust, perhaps solar systems in the early stages of development. A British scientist named James Jeans thought that perhaps the spiral nebulae were new matter, being introduced into the universe from another dimension. He was ahead of his time!

The controversy was settled by two men named Edwin Hubble and Milton Humason. Working with the 100-inch telescope on Mount Wilson, Hubble and Humason noticed that the spirals contained stars that fluctuated in brightness on a regular basis. Such stars, called Cepheid variable stars, were already known to exist in our galaxy. Some of the spiral nebulae contained several such variable stars! This seemed to indicate that the spirals were, in fact, enormously distant groups of stars. Hubble and Humason found that there was a strong correlation between the brightness of a Cepheid variable and its period, or length of time between peaks of brilliance. When this function was applied to the Cepheid variable stars in the spiral nebulae, the result did not come as a great shock to these astronomers. The spirals were, in fact, very far away. Some of the nebulae checked by Hubble and Humason were as distant as 10 million light years. Beyond that distance, the Cepheid variable stars were not bright enough to be seen and checked with accuracy. But many spirals were much fainter than those that Hubble and Humason saw at a distance of 10 million light years. These island universes were called galaxies. We know now that our own galaxy is one such spiral. There are several other galaxies relatively near ours. The whole cluster, including our Milky Way galaxy, is called the local group. See Figs. 1-12, 1-13, and 1-14.

Hubble and Humason used the brightness of each spiral, taken as a whole, to estimate the distances to far-off galaxies. Other types of galaxies were found to exist: Some were elliptical or spherical, and some were irregular in shape. Eventually, galaxies several billion light years away were photographed by astronomers. Some of these galaxies are hardly distinguishable, to the inexperienced eye, from ordinary stars. Their large red shifts, however, give them away.

THE HUBBLE CONSTANT

Hubble and Humason measured the distances to many different galaxies, and found that the amount of red shift was correlated with the distance. This correlation was astonishingly linear, as shown in Fig. 1-15. For each billion light years of distance to a galaxy, its speed away from us appears to increase by 30 millions miles per hour. Figure 1-15 illustrates a very recent estimate of the speed-versus-distance relation. The original estimates by Hubble and Humason were somewhat smaller than this. The slope of the line in the graph of Fig. 1-15 is called the Hubble constant. This value has been revised several times since the first such graph was plotted by Hubble. The Hubble constant tells us how rapidly the universe is expanding, provided, of course, that the red shifts actually are the result of expansion, and not some other unknown cause.

Observations of many galaxies have shown that the Hubble constant is the same in all directions. Evidently, all the distant galaxies are moving away from us at a speed that does not depend on their direction. Does this mean that we, in our Milky Way, occupy the center of the universe?

Hubble pointed out that, if all the galaxies were retreating from all the others in a uniform manner, then any observer, anywhere in the universe, would obtain the same value of 30 million miles an hour per billion light years as the Hubble constant. This is a simple mathematical truth; it can be demonstrated in a variety of ways. A violent explosion throws particles outward at various different speeds. The most rapidly moving pieces travel the greatest distance per unit time, and therefore they are always at the outer edge of the explosion. Slow-moving particles are always nearer the center of the explosion. If we could ride along on any one piece, we would see all the other particles retreating from us. The farther away they were, the greater their speed of retreat would be.

Fig. 1-12. Our galaxy, the Milky Way, is a spiral. It probably resembles the spiral galaxy shown here, which is NGC224 in the constellation of Andromeda (courtesy of Mount Wilson and Las Campanas Observatories, Carnegie Institution of Washington).

Fig. 1-13. This beautiful spiral galaxy is NGC5457 in the constellation of Ursa Major (courtesy of Palomar Observatory, California Institute of Technology).

Fig. 1-14. This spiral galaxy is NGC5194 in the constellation Canes Venatici. The satellite galaxy at the bottom is NGC5195 (courtesy of Palomar Observatory, California, Institute of Technology).

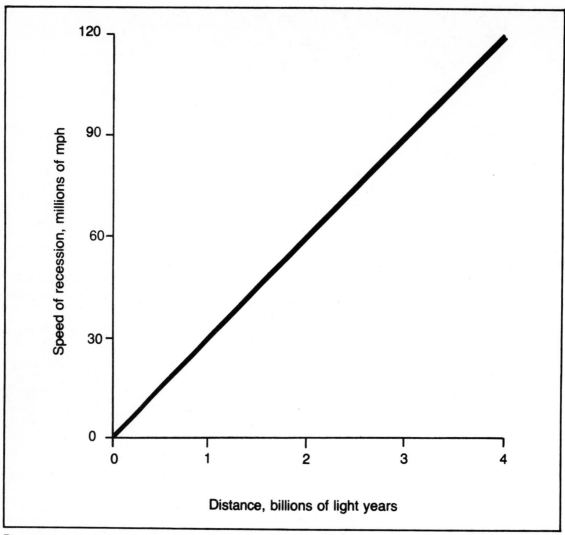

Fig. 1-15. In the twentieth century, Edwin Hubble found that the speed of the distant galaxies, directly away from us, is related to their distance according to this function.

We might get the feeling, as we rode on one particle, that we were at the location of the original explosion. But it would not necessarily be so.

It was at the time of Hubble's discovery that cosmologists first got the idea that our universe might, itself, have come from a great explosion.

THE EXPANDING UNIVERSE

When Albert Einstein first put forth his idea of a finite but unbounded universe, he thought that the cosmos must look the same from all points of view. If this was true, he reasoned, then the universe should look the same at all times as well. Einstein, and many other scientists, believed that the universe never changes. The discovery of the red shifts in the spectra of the distant galaxies, and the fact that this red shift depends on the distance, upset this idea.

Willem de Sitter found that Einstein's equations, which Einstein had formulated to describe the structure of the cosmos, contained a solution that implied an expanding universe. This discovery was made independently of the empirical observations of Slipher, Hubble, and Humason. Einstein had not noticed this solution to his equations. A Russian astronomer named Alexander Friedmann, independently from de Sitter, also found that the solution to Einstein's equations could lead to the idea of an expanding universe. This time, it turned out, Einstein had made a fundamental error in his calculations: He had divided both sides of his equation by an expression that could, under certain conditions, be equal to zero. Such a mistake, as anyone acquainted with algebra knows, can sometimes produce false solutions, and this had happened to Einstein, making him conclude that the universe was static. Initially, Einstein did not believe that Friedmann or de Sitter could be right. But finally, even Einstein admitted that the universe was, apparently, expanding according to his own equations.

Nevertheless, the idea of an expanding universe bothered Einstein, for it implies that all the matter in the cosmos was, at some time past, in one place. A creation! Many scientists were greatly disturbed at this thought. It has theological implications, for a cosmic beginning might have only one explanation: God created it! Science and religion seemed headed for an inevitable rendezvous. But, in the continual search for knowledge, nothing is more dreaded than the chance that it might actually all be found.

THE BIG-BANG THEORY

If everything in the universe is moving away from everything else on a large scale—and there is strong evidence for this—then evidently the universe was much smaller in the distant past than it is today. If we trace far enough back in time, perhaps there was, indeed, a tremendous explosion. This idea, called the big-bang theory, is the most popular concept of the universe at the present time. The universe, according to the big-bang theory, began

as an intensely bright and hot fireball, concentrated in a single point of space.

As soon as scientists got the idea that there had been a great explosion marking the beginning of the universe, two more questions immediately came to mind. First, when did it happen? And second, where did the original matter come from? The second question is still a great mystery, although some fantastic ideas are being pursued. The first question is answered by the value of the Hubble constant.

For each billion light years of separation, the speed of two cosmic objects, with respect to each other, is about 30 million miles per hour. This value is not known with great accuracy, but the value currently known represents about 5 percent of the speed of light for each billion light years of separation. At a separation of 20 billion light years, then, two cosmic objects ought to be moving apart at the speed of light. The age of the universe, based on the Hubble constant, is therefore 20 billion years.

An interesting thing about the Hubble constant is that it gives us, in addition to the age of the universe, a clue to its size, as well. If galaxies at a distance of, say, 20 billion light years are receding at the speed of light, then we cannot possibly hope to see anything farther away than 20 billion light years. Moreover, anything farther away than that would have to be moving at a speed faster than light, which is not possible according to the theory of relativity. This imposes a limit on the radius of the universe!

When we peer out into the cosmos, we are looking into the past. At a distance of 20 billion light years, give or take a few billion light years to account for possible errors in the measurement of the Hubble constant, we are looking at the very beginning of things. At that distance, then, what would we see? Perhaps we would see the glow of the cosmic fireball, just as it exploded. What would that look like? In 1948, two scientists named Ralph Alpher and Robert Herman had an idea about it. The residual radiation from the fireball would be greatly red-shifted, of course, and it would come from all directions simultaneously. If such radiation were observed, then the big-bang theory would become far more plausible as an explanation for the structure and evolution of the universe.

But there was a great amount of controversy around the time of Alpher and Herman. Many scientists disliked the big-bang theory because it suggested there was a creation, and it did not offer any concrete explanation of what happened before the cosmic explosion. So many scientists tried to shoot the big-bang theory down. Any really great theory must, sooner or later, pass this kind of test, anyway. The big-bang theory has survived all challenges. Nobody has come up with a theory that works any better.

THE STEADY-STATE THEORY

Einstein, you will recall, originally believed in a static or unchanging cosmos. He was eventually convinced by Friedmann and de Sitter to change his

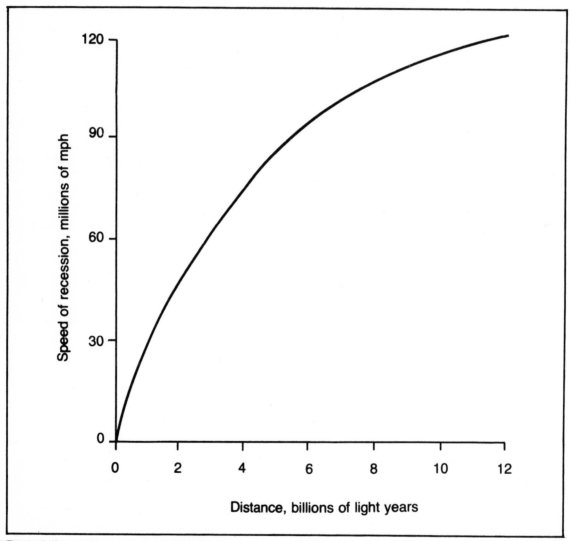

Fig. 1-16. If the universe was expanding more slowly in the past than at present, the speed of recession of the distant galaxies would increase with distance in this manner.

mind. But even if we accept the concept of an expanding universe, we can avoid the creation idea by theorizing that the rate of expansion was slower in the past than it is now. If this is true then the Hubble constant is not really a constant; the function should be a curve rather than a straight line, allowing much greater extensions into the past (Fig. 1-16). This would involve only a miniscule change in the value of the function for the more nearby galaxies; the distances and radial speeds of the far-off galaxies would be called into question. That is not hard to do when we remember that the value of the constant is rather uncertain anyhow, and has been revised several times since its first discovery by Hubble.

A Cambridge University student named Tommy Gold, along with Herman Bondi, one of his professors, postulated that the universe has always existed and always will. The astronomer Fred Hoyle joined Bondi and Gold to formulate what is now called the steady-state cosmology.

According to the steady-state model of the cosmos, the density of the universe must remain the same at all times. Not only will the universe appear the same no matter from where we look at it, but it must always seem the same no matter *when* we look at it. How can this idea be made consistent with the idea that the universe is expanding? Well, obviously, new matter must be coming into existence throughout the universe, and its rate of formation must be just enough to offset the expansion. Gold, Bondi, and Hoyle suggested that new matter, in the form of hydrogen atoms and appearing from nowhere, was forming all over the universe at the rate of one atom per million cubic meters per century. This appears to contradict the law of conservation of mass, but on a cosmic scale, this law might not be valid.

The steady-state idea, thus formulated, was the focus of a great controversy in the scientific community. From where would all of these new atoms come? The creation problem had resurfaced! It seemed that there was no escape from that idea. The new matter might be introduced from another dimension, perhaps through "holes" in space. Or, it might be created very quietly, out of nothing.

Perhaps it could be made from energy, already present in the universe. Einstein has shown that matter and energy are actually manifestations of the same thing, and are related in his relativity theory by the famous equation $E = mc^2$. We know already that matter can be converted to energy; this happens in every star. Perhaps energy is converted back into matter!

The big-bang and steady-state theories are the only two contemporary models of the universe that have been taken seriously. But which, if either, model is correct?

We now believe that the big-bang theory is more plausible than the steady-state theory. And the reason is one more bit of experimental data, obtained with the help of the radio telescope.

ECHO FROM THE FIREBALL

If the big-bang model of the universe is correct, then it ought to be possible to actually see the cosmic explosion if we look far enough out into space. Alpher and Herman predicted that it would be found. But we must use radio receivers to find it, because of the extreme red shift involved.

Imagine probing into the depths of space in some particular direction, such as toward the north celestial pole. As we peer far beyond Polaris, the North Star, to a distance sufficiently far off, the red shift becomes greater and greater. Eventually the corresponding radial velocity will become almost the speed of light. Finally, there will be a distance beyond which we cannot see, because all light will have been red-shifted to a frequency of zero.

Suppose we now turn our optical and radio telescopes toward the constellation Orion, near the celestial equator. Again, we will observe greater and greater red shifts as we probe into the depths. The red shift will, again, become total if we look far enough. We will see the same thing in this part of the sky as we do toward the north celestial pole. In fact, no matter what part of the heavens we examine, the same thing will happen.

At a radius of several billion light years around the earth, in a gigantic celestial sphere, lies the edge of the observable universe. It is also the edge of the conceivable universe, for at this distance, relativistic time dilation renders everything at a complete standstill. The objects at the edge of this sphere are hurtling away from us at almost the speed of light! This great black sphere represents the universe at some distant time past—between about 15 billion and 25 billion years, depending upon the precise value of the Hubble constant. But the sphere is not quite totally black!

In 1965, Arno Penzias and Robert Wilson were working in the Bell Laboratories with a sensitive radio receiver when they noticed a peculiar background noise. They thought, at first, that something was wrong with their equipment, and that the noise was generated inside their own receiver circuitry. But a careful check showed that the noise was external in origin. Not only that, but it was not even coming from the earth! Strangest of all, the noise appeared to come from everywhere in space with equal intensity.

Investigations of the nature of this noise were conducted. It was found that the detectable wavelengths extended over a wide range. The peak intensity was found to occur at about 0.3 centimeters. Instruments were put aboard a jet airplane and flown to a high altitude to get rid of most of the interference generated on the earth; the radiation was checked from a balloon at an even greater altitude. Scientists at Princeton University in New Jersey, and at Cambridge University in England, measured the intensity of the radiation at various different frequencies. When the data were assembled, there could be no mistake: The relation between signal intensity and wavelength was definitely the sort of curve characteristic of an explosion. The cosmic radiation corresponds to a temperature of about 3 degrees Kelvin. This is the extent to which the cosmic fireball has cooled since it first blew up. The big-bang theory is apparently correct! Scientists had predicted that the temperature of the fireball would be about 5 to 10 degrees Kelvin—remarkably close to the value actually found.

Imagine, if you can, the matter of all the universe—millions of galaxies, hydrogen gas clouds, black holes, quasars, and unknown objects—crammed into a tiny space no larger than a subatomic particle! Where did this great particle, which cosmologist George Gamow has called ylem, the ancient name for a substance from which all things are made, come from? Why did it explode? We cannot be certain about these things, but some fantastic ideas are being formulated about it. Scientists are beginning to accept the idea that a creation actually did take place. Creation may be a frequent occurrence in a space-time continuum that more nearly resembles Swiss cheese than Newton's smooth-flowing milk. Perhaps creation occurs with infinite frequency. We must begin to open our minds to the extent that we can imagine different dimensions of time as well as of space. We must acknowledge that perhaps our beginning was preceded by someone else's doomsday, and that the infinite hereafter in our cosmos may herald the birth of another universe.

THE END OF THE UNIVERSE

We have spoken a great deal, so far, about beginnings and creation, and have said nothing about endings and destruction. Is this simply optimism? No. Sometime the universe might well come to an end, just as it apparently came into being. How might the end of the universe arrive? There are two theories, which might be called the ice theory and the fire theory in a descriptive sense.

In deciding whether the end will be cold and dark, or hot and bright, the primary question is this: Will the universe ever fall back together again? Will the gravitation among all the matter in the cosmos be sufficient to reverse the outward motion of the galaxies? Or is the radial velocity too great? In order to answer this question, we must know how much matter there is in the entire universe, or at least what the density of the matter is.

The visible stars and galaxies do not account for all of the matter in space. Much of the mass of the universe is comprised of invisible gas and dust between the stars. In fact, this tenuous material, which we cannot see and may actually block our view of the visible universe to a large extent, probably accounts for most of the matter in space. Dead stars, having spent the last of their atomic energy, emit no light; but they still can be as massive as ordinary stars. Some invisible objects, called black holes because their extreme gravitation literally rips them away from space and time, could perhaps account for much of the mass in our universe. Any estimate of the density of matter in the cosmos must, of necessity, be a rough, educated guess, and it could be off by a large factor.

Calculations by cosmologists indicate that, if the universe contains approximately 35 or more hydrogen atoms (or the equivalent amount of mass) within ten cubic meters, on the average, then the gravitational force produced by all the matter in the universe ought to bring the expansion to a halt, and someday reverse it, causing contraction.

The best estimates of the density of matter in the universe seem to indicate that there is not enough to cause the expansion to stop. In fact, the density of matter in the cosmos appears to be less than 10 percent of the required value. Apparently, then, the universe will keep on expanding, and all the stars will eventually die and become dark. There is plenty of hydrogen left in the universe, at the present time, to keep it going for a long time, but there is not an unlimited supply. Although it may take hundreds of billions of years, the universe will probably come to a cold, dark end, unless something is introduced to keep it going. This, then, is the ice theory.

But what if our estimates of the density of matter in the universe are too small? What if the educated guess is off by an order of magnitude? Then the gravitation among all the atoms in the cosmos will be sufficient to pull the universe back together again. This will at first be manifested as a reduction in the rate of expansion. Were we to look at the distant galaxies as the universe contracted, we would see not a red shift, but a blue shift—a shortening of the wavelengths in all the spectra of the spiral nebulae. As the universe continued to collapse, its inward velocity would become greater and greater. Finally, all the dead stars, the black holes, the remaining gas and dust, would fly inward and disappear into a brilliant, hot, tiny ball of light. Time and space would again be squeezed into a geometric point, which cosmologists call a space-time singularity. This is the fire theory!

The fire theory is more emotionally satisfying, to be sure, than the ice theory. The fire theory allows for the possibility that our universe oscillates—that is, it recurs in an alternating sequence of expansions and contractions. More research will certainly be carried out in the coming years, in an effort to determine the true density of our universe.

AN OSCILLATING UNIVERSE

If our universe actually does contain enough mass to halt and then reverse the expansion, what will happen during those final moments, as all the material in the cosmos comes crashing down upon itself? The gravitational force in such a collapse would be immense—unlike any other force in the universe. The gravitational field of a collapsing star is intense enough to prevent even light from escaping, as we shall see when we look at the mysteries of the black hole. Compared to the pull of an entire universe concentrated into a volume the size of an atomic nucleus, however, the effects of a black hole would seem trivial. The matter of the whole universe might even be squeezed into a single geometric point!

There are two theories concerning the events following a final collapse of the universe. The first idea is that the matter in the cosmos will simply stay packed together forever. If that is the case, it would be the end of the universe, and the finality would be as irreversible as that of the ice theory. The only difference from the doomsday of the ice theory would be the temperature, which would be trillions of degrees instead of practically absolute zero, and the density, which would be incredibly great rather than practically zero.

A more appealing idea, since it provides some hope that the universe will not face an ultimate demise, is the theory that the matter might rebound. This would create another cosmic explosion, and send the matter flying outward again! Another universe would be born. The primordial fireball would again cool and disperse, and perhaps atoms, stars, and galaxies would again condense from the ylem, that strange fundamental substance from which all matter comes. Such a universe might be very much like this one, or it might be utterly different. Perhaps the atoms would have the same structure, or perhaps they would have much different structures. Maybe the very laws of physics, as we know them, would be greatly altered by the supernatural cataclysm of collapse and rebirth. We do not know.

As the old universe collapsed, so, eventually, might the new. The laws of physics might be the same in the next universe as in this one, especially the laws of gravitation. The expansion would then continue only until the new universe reached a certain maximum size. Then the inevitable contraction would begin, just as in the previous universe, and the inhabitants of some tiny planet, orbiting some star in some galaxy, might notice a blue shift in their spectroscopes. The volume of their universe would become smaller and smaller over many billions of years. Finally, with all the violence of the primordial explosion, the new universe, too, would end. All the elementary particles of the new universe, whether the same as those of other universes or not, would again be crushed into ylem, the fundamental substance. And, as before and always, the matter would rebound and form another universe, perhaps with the same properties, and perhaps with different properties.

This is a beautiful theory, and has been called the oscillating-universe model. Such a theory gives us something to look forward to, even after the end of all things. You might wonder why we should be concerned about this, and why we should try to extend our horizons beyond a trillion years when there is some question as to whether we will exist for ten more years. Isn't a few hundred billion years enough time for our universe? Is it not quite ridiculous to attempt to justify an eternity?

Such cynicism does not occur to cosmologists who study the universe. The very idea that the cosmos might last forever is testimony to the great power of things. Even the most objective scientist cannot help but feel a little bothered at the notion that time and space might actually come to an end. It is human nature to think about, and believe in, forever!

BEYOND SPACE

Einstein, as we have mentioned, was one of the first, to put forth the idea that space might be curved, and that the universe is finite but unbounded. Such strange ideas as spatial curvature and a spherical universe tug at the limits of our imaginations. Even today, some people cannot accept such ideas. We cannot readily visualize more than three spatial dimensions. What would it be like to be a four-dimensional creature? Imagine being able to take four yardsticks and place the ends together at a single point, and then arrange the yardsticks so that each one is perpendicular to the other three! It is too much to think about. But mathematically, it can be done in a simple and elegant way.

According to Einstein, our universe is like a bubble in four-dimensional space. We, along with all the stars, planets, galaxies, and other objects in the universe, exist on the three-dimensional surface of this sphere. To get some idea of our situation, we can imagine a geometric sphere of the ordinary three-dimensional kind that is familiar to us. The surface of such a sphere is two-dimensional.

Do you recall, from your childhood days, blowing soap bubbles from an apparatus designed especially for that purpose? Suppose, for a moment, that each of those bubbles was an entire universe! Imagine that, on the surface of each bubble, two-dimensional stars and galaxies evolved. Suppose that some of the stars had planetary systems, and that on a few of these planets—maybe a million or so for each bubble—intelligent civilizations grew. Suppose that the lives of the soap bubbles, just a few seconds to you, seemed billions of years to the inhabitants of those planets. The universe of the soap-bubble surface, having just two dimensions, would allow its inhabitants less freedom of movement than we, in three-space, are accustomed to. Two-space has been termed "flatland." But it is easy to see that the surface of a soap bubble is not really flat. In a local sense, it is nearly flat. Over an area of a millimeter by a millimeter, for example, the surface of the average soap bubble is flat for all practical purposes. The solar systems of the inhabitants of the soap-bubble universe might only be about the size of our atoms. Therefore, the spherical nature of their universe might well escape the attention of such beings.

How would the curvature of the soap-bubble universe be detected by beings in two-space? Well, their mathematicians would surely have noticed that the sum of the measures of the interior angles of any triangle adds up to 180 degrees. They would certainly have invented their own Euclidean geometry. And with just two dimensions to worry about, their geometry classes would be a great deal easier than ours! A fairly advanced two-dimensional civilization might invent telescopes, spectroscopes, and lasers, and begin to investigate the puzzles of gravitation and cosmology. Someone might, using three lasers, three receptors, and a precise angle-measuring system, challenge the validity of the 180-degree rule for triangles. With accurate

enough equipment, a slight deviation from this rule might be noticed.

On the surface of a sphere, the measures of the interior angles of a triangle do not add up to exactly 180 degrees (Fig. 1-17). Instead, the sum is larger than this value. The difference is tiny when the triangle is small by comparison to the whole sphere. As the triangle is made larger, the difference increases. You can verify this with three pieces of string, some tape, and a globe or basketball. A triangle with one vertex at the north pole and two vertices on the equator, when the equatorial vertices differ by 90 degrees of longitude, has three 90-degree interior angles, with a total internal angle measure of 270 degrees. In the extreme case, a triangle on a sphere might have three vertices which all lie along the equator. Then, each internal angle is 180 degrees in measure, and the total is 540 degrees!

The contradiction having thus been found, the

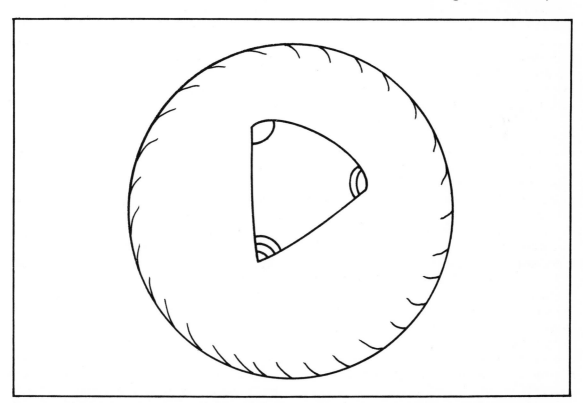

Fig. 1-17. On the surface of a sphere, the sum of the interior angles of a triangle adds up to more than 180 degrees.

two-space creatures would have to throw away their ingrained notion that their universe was flat. *Reductio ad absurdum* must hold just as well in two dimensions as in three!

It is possible that the two-space beings, if especially imaginative, might get the notion of curved space without having to conduct experiments. Perhaps some brilliant theoretician would derive, mathematically, that result, based purely on the elementary laws of gravitation. In our three-space universe, this is just the way that it happened. The theoretician's name was Albert Einstein, and he discovered the curvature of space early in the twentieth century.

If our universe is really like a soap bubble in four-space, then the question naturally arises: What else is there in four-space? We have no way of knowing. Our three-dimensional continuum is infinitely thin in such four-space. The soap-bubble analogy is not quite perfectly accurate, then, because even a soap film has some thickness. With respect to four-space, our universe has no thickness at all. A two-space, relative to our universe, would be infinitely thin, as well.

Our imaginations, then, are the limit! Perhaps our universe is one great four-dimensional accident. Or maybe it is some laboratory experiment, created with four-dimensional electromagnetic fields. Perhaps the technicians who created our universe, knowingly or unknowingly, are about to get off from work and go home. They reach for the power switch . . . Or maybe it's only their lunch hour!

BEYOND TIME

What is time? Astronomers, cosmologists, philosophers, and mathematicians have wrestled with this question for centuries. Isaac Newton thought that time flowed smoothly, and always at the same uniform rate. Some scientists have treated time as a dimension, just like the spatial dimensions with which we are familiar. Einstein was one of the first to treat time as a dimension, and he considered it to exist perpendicular to all of the spatial dimensions. Some important results were derived from Einstein's model of time. By treating time as a dimension, we can gain some understanding of four-space.

In a two-dimensional universe, a circle, square, or other closed plane figure would be sufficient to imprison someone. We, in three dimensions, can easily imagine how such an imprisoned creature might escape: It would only be necessary to lift him out of the two-space universe and set him back down outside of the closed figure. With access to four dimensions, any of us could escape from a sealed box with equal ease. One method of gaining access to the fourth dimension would be to travel back and forth freely in time. We could then propel ourselves into the future until the box decayed or was dismantled. We could then simply step aside a bit, and return to the present to be outside the box! Alternatively, we might travel back in time until before the box was built, step over a few feet, and return to the present.

Einstein carried the dimensional aspect of time still one step further than this. Time, Einstein believed, was essentially no different than the dimensions we perceive as spatial. Time, according to Einstein, was simply a manifestation of space; time and space are fundamentally related by the speed of light. You have surely heard the expression, "The store is ten minutes away from our house," or something similar. A given distance, in absolute terms, is simply equal to time multiplied by the speed of light. A light second is just that: One second multiplied by the speed of light, or 186,282 miles.

This notion gives rise, in conjunction with the theory of a finite but unbounded space, to one of the most bizarre ideas imaginable. If traveling in one direction for a great enough distance will bring us back to our starting point, then perhaps the future will eventually become the past, and then the present! Perhaps time, like space, is finite but unbounded!

All of us have seen, or drawn, time lines. In high-school history class, for example, the teacher might have drawn a time line extending from about the year 5,000 BC to 2,000 AD, marking approximately the extent of the history of known civilization. This 7,000-year span is but a heartbeat in the history of our planet. And it is an even tinier instant in the perspective of the universe; in comparison to the elapsed time since the big bang, man has been civilized for almost no time at all. But, although the big bang certainly took place a long while ago, time might not be infinite in extent. Suppose that the ultimate time line is finite but unbounded—that is, circular! Figure 1-18 shows what a complete time-circle model looks like, consistent with the idea of an oscillating universe.

The big bang is at the top of the time circle, corresponding to 0000 universe hours on a 24-hour system, similar to that used by the military. The big bang thus took place, in this time model, at universe midnight. Time proceeds clockwise around this circle. Between 0000 and 1200 hours, representing the universe morning, the cosmos expands. The universe reaches its maximum size at 1200 hours; in the cosmic afternoon, it begins to contract. The rate of contraction increases as the hour gets later. Finally, at 2400 universe hours, the universe falls back to the primordial state. We are presently living in the small hours of the universe morning—say about 0200.

If the oscillating-universe theory is in fact true—and we do not know that it is—then the

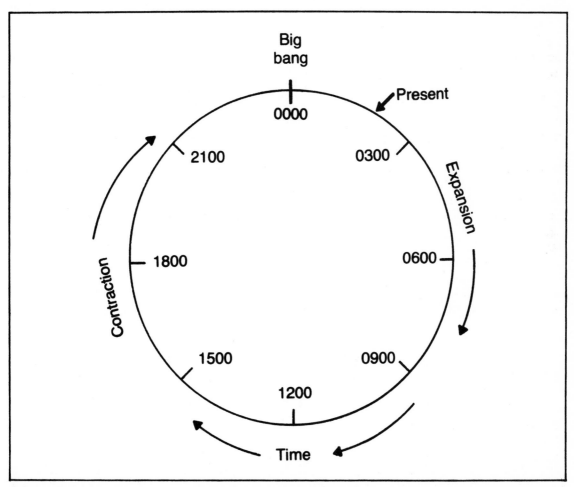

Fig. 1-18. The universe time clock, according to the model of an oscillating universe, begins with the big bang and ends with collapse. But after the collapse, another day begins. In such a model, we are, at present, living in the small hours of the morning.

question arises: Is each universe day unique? Will the next cycle be much different from this one? Or will it be similar, or perhaps even *identical*? We will probably never know the answer to this question, even if we someday find that the universe has sufficient mass to bring about an eventual collapse. But we can be fairly sure of one thing: The truth about our universe is probably more strange than the wildest science fiction.

As you rest on a hilltop and view the sky on a clear, moonless night, you are gazing at the same stars and galaxies that have fascinated philosophers and scientists since the dawn of man. Modern researchers have complex and precision tools to use in the pursuit of the secrets of the cosmos. Ancient men had only their eyes. The first men, as they looked at the sky, knew little; there was one question and no answer. Today there are perhaps infinitely many questions, and a few answers. How far have we come? In relation to the whole, we have scarcely moved a step. But the same vision still drives us.

Chapter 2

The Sun,
Stars, and Galaxies

I T IS NOT NECESSARY TO SEARCH HUNDREDS OF millions of light years away from our solar system in order to encounter the mysteries of the cosmos. Some of the most unfathomable puzzles are comparatively close to us. Our own sun operates according to a principle that astronomers and physicists still do not fully understand. The formation and evolution of the solar system is not completely known. Many stars exhibit strange behavior, which is difficult for us, with our plain, ordinary sun, to readily imagine. Our own galaxy, the Milky Way, is very hard to study in many respects, oddly enough because of its very proximity. We are right in it—it is hard to see from such a vantage point!

Of course, there is no shortage of mysterious phenomena outside our Milky Way galaxy. Why do some galaxies look so much different than others? Why do some seem to be spinning around, while others just hang there in an irregular mass? How far away from us do the galaxies extend, and what is between them? And then, of course, there are the recently discovered quasars; astronomers are not

in complete agreement as to what these objects are.

In the search to answer for cosmological questions, astronomers have invented different kinds of telescopes. The earliest telescopes, operating only in the visible-light portion of the electromagnetic spectrum, give us only one picture of universe with many faces. Space objects have a profile at all wavelengths, not just the ones we can see with our eyes. This is true from the lowest radio frequencies, with wavelengths of hundreds of kilometers, to the shortest gamma rays, with wavelengths only a tiny fraction of a millimeter.

At all electromagnetic wavelengths, one of the simplest celestial objects to study, because of its proximity to us, is our sun. Only within the last century have scientists begun to really understand what the sun is, and how it works.

WHAT IS THE SUN?

Even the ancients recognized two things about the sun: It is very bright, and it is very hot. Until recently, however, physicists did not know how the sun operates. In ancient and medieval times, the sun was thought by most to be relatively small and nearby. There were a few exceptions, but no one got the idea that the sun might be as remote and huge as it actually is. Some people believed that the sun was no more than a few feet across, and that its path above the earth was at an altitude of only a few miles.

You may have heard the legend of Icarus and Daedalus, who flew high above the earth by using artificially constructed wings. The wings were made from feathers, and held together by wax. Daedalus knew that the sun was hot, and not very far up; he avoided excessive altitude because he knew that the wax in his wings, holding the feathers intact, would melt if it got too hot. Icarus was more foolhardy. He flew, in his exuberance, too high, and too close to the sun. The heat of the sun melted the wax in Icarus' wings, and he fell into the sea and was drowned. Today, of course, we know his wings would have been in no danger; he probably would have frozen to death instead! We know that the sun is so far away that, at an altitude of several miles, the distance to our parent star is essentially the same as from the ground.

The ancient astronomer and mathematician, Eratosthenes, was on the right track when he assumed that the sun was so far away that its position in the sky would not change when seen from different places on the surface of the earth. Eratosthenes, who believed that the earth was spherical, used this assumption to obtain a surprisingly accurate estimate of the circumference of our planet. He was far ahead of his time. Even today, some people reject the findings of Eratosthenes, and insist that the earth is flat.

It would be easy to determine the distance to the sun if we knew its diameter. Conversely, determining the size of the sun would be fairly simple if we knew how far away it is. But we must know one of these variables before we can determine the value of the other by trigonometry. We do know, and the ancients noticed as well, that the sun and the moon are almost exactly the same apparent size in the sky. Both objects subtend about ½ degree of

arc. But, of course, the moon is much smaller and nearer than the sun.

How can we find the distance to, or the diameter of, the sun? The distance to the moon can be determined by looking at it from two different places on the earth, and noting the difference in position with respect to the background of distant stars. This is called triangulation. From New York City, for example, the moon appears further south in the sky than from Buenos Aires. By measuring the angular distance that the moon moves as we go from New York to Buenos Aires, and knowing the base line between the two cities, the distance of the moon can be found by triangulation. The same thing can be done with the sun, although the instruments required are more sophisticated than those needed to measure the parallax of the moon. We know today, of course, that the sun is about 93 million miles from the earth, and that our distance from the sun varies by about a percent either way from this value. We know that the sun is about 864,000 miles in diameter, or over 100 times the diameter of the earth.

Using Newton's laws of gravitation, and knowing the length of time required for the earth to orbit the sun, we can determine the mass of the sun. It turns out the sun has about 300,000 times the mass of the earth, but the sun is less dense than our planet—about 1.4 times the density of water.

Astronomers before the twentieth century surmised that, with all that material in the sun, the gravitation had to be tremendous! The compression of matter resulting from such a powerful gravitational field could, they thought, perhaps account for the energy output of the sun. We all know that compression makes things get hot. But when calculations were carried out to check the theoretical amount of heat and light that should be produced, the result fell far short of the actual energy from our sun. This model, then, could not adequately explain things. The difference was unexplainable for some time.

Then how, physicists wondered, does the sun work? This question remained unanswered until Albert Einstein showed that elements could be changed from one to another, and that this conversion would often result in the liberation of great amounts of energy. We have seen the simple, but profound, summarization of this principle: the formula $E = mc^2$. This formula is simply an expression of the relation between matter and energy. Einstein said that matter and energy were the same thing, but in different forms. Given an amount of mass m, it can be converted to a quantity of energy E, and the relation is given by the formula. Conversely, a certain amount of energy E is equivalent to a mass m according to the Einstein formula. The factor c^2 is the speed of light multiplied by itself, in the system of units chosen.

THE SOLAR MODUS OPERANDI

Physicists of today believe that the sun converts hydrogen to helium in a vast atomic reaction, called nuclear fusion. This principle is the same as that which makes a hydrogen bomb explode. But other elements, not just hydrogen and helium, are present in the sun. There are even some compounds—combinations of more than one element, joined together.

Hydrogen fusion requires extremely high temperatures. Such temperatures can be attained by compression of the sun by its own force of gravitation. At first, two hydrogen atoms combine to form one atom of deuterium. Hydrogen is the simplest existing element, containing one proton and one electron. Deuterium consists of one proton, one neutron, and one electron. Two protons can be squeezed together, and emit a particle called a positron, or anti-electron; the result is a nucleus of deuterium, or heavy hydrogen. This nucleus can combine with another proton to form a nucleus of helium 3, containing two protons and one neutron. Two of these nuclei can then combine to form a nucleus of helium 4, which has two protons and two neutrons. In this process, two protons are given off. Helium 4 is the form of helium with which we are familiar. (We use it to fill lighter-than-air balloons and dirigibles.) The end result of all this nuclear juggling is that four protons, or hydrogen nuclei, get

transformed into a single nucleus of helium. This is shown in Fig. 2-1.

A proton has a mass of 1.008 atomic mass units. So we would expect that four protons, having a combined mass of 4.032 atomic mass units, should have the same weight as a helium nucleus. But this is not the case; a nucleus of helium 4 has a mass of only 4.003 atomic mass units. The remainder of the mass—0.029 atomic mass unit, or 4.8×10^{-26} gram —has to be accounted for. But how? The answer is that it gets converted into energy. This happens according to Einstein's famous equation. Each second, the sun produces a tremendous amount of energy in this way. There is enough matter in the sun to keep this process going for

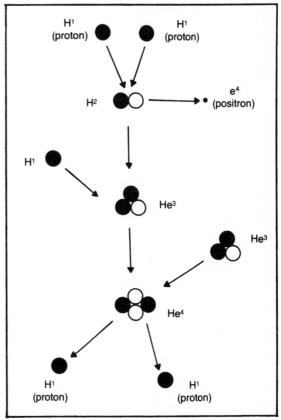

Fig. 2-1. It is theorized that, inside the sun, hydrogen atoms combine to form helium atoms by means of nuclear fusion. In this chain of events, tremendous energy is released.

billions of years. The nuclear-fusion model explains, very elegantly, how the sun can be as bright and hot as it is.

Nevertheless, there are still some problems with this model. The discovery of the neutrino, a strange cosmic wisp that can travel through almost anything unimpeded, has cast an aura of mystery around the operation of the sun. While it is still believed that nuclear fusion accounts for the energy from the sun, this reaction may not be going on in a continuous manner. The solar furnace may shut down and then restart periodically, at intervals of millions of years, in a manner very much like the furnace thermostat mechanism in a home! This might, in turn, produce fluctuations in the light and heat coming to us from our parent star, whose stability we take for granted.

THE NEUTRINO QUESTION

Physicists who study atoms and atomic particles have no shortage of strange phenomena in their field. There are all kinds of different particles of matter and energy. New types of particles are constantly being found. Particles may be discovered in either of two ways: theoretically or experimentally.

When the existence of a certain kind of particle, having certain properties, is predicted by theory, the experiments begin in an effort to search for the effects of this particle. Sometimes, during the experimentation process, certain things happen that aren't expected. Then, the theorists must find an explanation for what is observed. One particle that was found by the theoretical method is the neutrino. Experiments have confirmed the existence of neutrinos. They are given off in nuclear reactions. But the results of the experiments have generated another puzzling question: A mysterious shortage of these strange particles has thrown a fascinating anomaly into our understanding of how the sun shines.

When we think of a subatomic particle, we always imagine something extremely small. Neutrinos fit nicely into that classification. We suppose that an atomic or subatomic particle has a certain amount of mass, albeit very small. Well, neutrinos have theoretically zero mass. Not just very tiny

mass, not only much smaller mass than any other particle, but no mass at all.

How can something even exist without having mass? If an infinite number of massless particles were congregated together, perhaps the whole group might have some mass. But any finite number of neutrinos—even a googol (10^{100}) of them—would still mass nothing. The neutrino travels at the speed of light, and has what physicists call a spin. That is intuitively tolerable. But the peculiar lack of mass, undeniably predicted by theory, gives the neutrino the astonishing ability to penetrate vast amounts of matter.

All cosmic objects appear essentially transparent to a stream of neutrinos. Our bodies, trees, buildings, mountains, and even the entire planet seem almost perfectly clear to the massless neutrino. Even whole stars offer very little resistance to them! The neutrino is the only known particle with this sort of penetrating power; all other kinds of radiation, including the most energic gamma rays, are stopped by just a few feet of earth.

Where do neutrinos come from? The theorists predict that they should be generated in the centers of all stars, including our sun. And since neutrinos have such penetrating power, they are able to escape into space from the deepest layers of a star. Theoretically, based on the mass of the sun and the amount of energy it generates, we should expect that about 9×10^{13}, or 90 trillion, neutrinos pass through every square foot of directly exposed space per second on the earth. If there is a way to detect these particles, then, there certainly shouldn't be any problem with scarcity! Furthermore, the intensity of neutrino radiation, in theory, hardly changes with the time of day. The particles can go through the earth more easily than light passes through a pane of window glass. At night, the neutrinos from the sun shine upward at us, from under the ground!

The extreme penetrating power of the neutrino is the main reason, however, for the difficulty in detecting this elusive particle. Very few neutrinos are stopped by any kind of obstruction. But a few is a lot more than none. Some of the particles are stopped, even by a comparatively small barrier. The common compound called carbon tetrachloride is affected by neutrinos. In the remote event that an atom of this substance, having the right isotope of chlorine, stops a neutrino, that atom will be converted into argon. Carbon tetrachloride is not a terribly rare or expensive substance. You might have used it as a cleaning agent, and recognized its alcohol-like aroma. According to the theoretical predictions, a massive tank of carbon tetrachloride ought to intercept several neutrinos each second, assuming that we are indeed being bombarded by the expected number of the particles. After a while, there should, then, be a small but measurable amount of argon in a tank of carbon tetrachloride.

The first experiments carried out in an effort to detect, and measure, the neutrino radiation in our vicinity were carried out in 1968. A huge tank of carbon tetrachloride was set up in an abandoned mine, far under the ground. Thus, the only particles that could reach the tank were neutrinos; all other particles would be stopped by the earth. Neutrinos were indeed found—but not in the quantity predicted. There were far too few of them. There were so few, in fact, that the chances of their originating in the sun were debatable. The theorists had to swallow the idea that the sun was radiating far fewer neutrinos than they knew it must; it might not be producing *any* neutrinos!

The experimentalists had done their job. Now it was up to the theorists to find an explanation for the results. According to the nuclear-fusion theory of solar radiation, neutrinos simply *must* be present, but they are not. The center of the sun, then, must be cooler than was originally surmised. But nuclear fusion is the only known means by which the sun can shine as we know it.

There appears to be one way out of this apparent impasse. Suppose that the nuclear furnace inside the sun goes on and off, at intervals of perhaps a few thousand or million years. Imagine that, at the moment (in a cosmic sense, that means a span of at least a few hundred thousand years), the furnace is off. The residual heat in the sun, left over from the previous fusion-reaction cycle, might keep it hot and shining. In fact, the cycling of the sun's great fusion reactor might be a sort of regulating device, intended by nature to keep things from getting out

of hand. Perhaps, if not for this regulating mechanism, the sun might actually overheat and explode! (Some stars do just that.) Then, if the sun is in an off cycle right now, we can be satisfied with the reality that there are very few, if any, neutrinos coming from its center.

This cycling of the solar furnace might have far-reaching effects on our own planet. The sun's thermostat, like any thermostat, is probably not absolutely perfect. A little bit of temperature variation no doubt takes place. The earth might warm up by several degrees, on the average, if the solar fusion reactor starts up again. There is some intriguing evidence that, sometime millions of years ago, the solar-furnace cycle was on, and the earth was considerably warmer than it is today.

SOLAR ACTIVITY AND THE ICE AGES

We have all heard of the dinosaurs, those great lizards that roamed the earth millions of years ago. There are a few descendants of these creatures still crawling around. But sometime in the geologic past, at a fairly well-defined time, most of the dinosaurs died out. The reason for this is a subject of fascination for scientists in both the physical and biological fields. The dinosaurs, at their zenith of evolution, had adapted to apparent perfection, and completely dominated the life system of our planet. They were ideally equipped to survive, even though they were not particularly intelligent. And then they just vanished. Why? There are many theories. One possibility is that the climate of the earth got cooler. Cold-blooded lizards would then find it very difficult to live, and they would eventually become extinct in most or all regions of the planet.

There is geological evidence that palm trees once grew at latitudes as far north as Greenland and Alaska. Then, the great ice age came. We are still living in that geologic winter. Sometimes the ice age grows bitter indeed, and glaciers form in the temperate latitudes. At other times, the ice age moderates somewhat. At the present time, we are in between bitter periods. These "cold fronts" and "warm fronts" seem to occur at intervals of several thousand years, but on the average, the climate of the earth is like one great long winter. When the dinosaurs reigned, the climate of our planet was like a great summer.

There are several theories that can adequately explain the short-term variations in our climate, but the long-term fluctuations, which apparently complete their cycles over periods of millions of years, might well be the result of variations in the radiation from the sun! This is an unsettling thought, but it is supported by the neutrino theory and the results of the experiments. We like to think of our sun as a constant. It is unpleasant to imagine that our parent star might be even a little erratic! But we know that bright light in the sky is far from the perfect, flawless yellow ball that the ancients believed it to be. The sun is covered with blemishes; it erupts in fits of rage at times, ruining radio communications on our little planet and creating the spectacular aurora; the sun spits out long red streamers thousands of miles into space. Is it so hard to believe that its intensity might change by a few percent in several million years?

Man need not, at any rate, fear the sun. The arrival of the next geologic summer will take thousands of years, at least; the more likely figure is millions of years. Certainly we can adapt to the change in that time.

ELEMENTS IN THE SUN

The earliest theories about the composition of the sun were based simply on the observation that it appeared to be burning. One idea held that the sun is a huge bonfire. Another theory presented the idea that the sun is a ball of petroleum. Today, we know that there are many chemical elements in the sun, and that the parent star is not a fire as we know fire. The most abundant elements are hydrogen and helium. But the spectroscope has allowed astronomers to detect more than 60 chemical elements in the sun. There are some compounds as well. All chemical substances produce radiation or absorption of certain energy wavelengths in the electromagnetic spectrum. Because of absorption in the solar atmosphere, the spectrum of the sun contains thousands of dark lines. But, to see these lines, the light from the sun must be passed through a narrow slit before it is put through a prism. A German optician named Fraunhofer was the first scientist to notice these dark bands, and he assigned them letters of the alphabet. Today, scientists still use these letters. They are known as Fraunhofer letters.

The wavelengths of visible light are extremely short, and are measured in units of 10^{-10} meter, known as Angstrom units. An Angstrom unit is so small that it cannot be seen with the unaided eye. Red light has the longest wavelength, in the neighborhood of 7000 Angstrom units. The wavelength gets progressively shorter through orange, yellow, green, blue, indigo, and violet. The shortest violet light that we can see has a wavelength of about 400 Angstroms. The position of a line in the spectrum is given in terms of its wavelength in Angstroms. Some of the most common absorption lines in the visible spectrum of the sun are the C line, at 6563 Angstroms, the F line, at 4861 Angstroms, the H line, at 3968 Angstroms, and the K line, at 3934 Angstroms. The C and F lines are the result of absorption by hydrogen. The H and K lines, which are quite close together in terms of wavelength and are therefore easily identified, are caused by calcium.

Oxygen is present in the sun, as indicated by the Fraunhofer A and B lines 7594 and 6867 Angstroms; these lines are enhanced in sharpness by absorption in our own atmosphere as sunlight filters through. Sodium is present in the sun; it shows the D line, at 5893 Angstroms. Close observation reveals that this line is actually two separate lines, extremely near each other in the spectrum. Iron is identified by the E line at 5270 Angstroms. These lines are all very conspicuous in the spectrum of the sun. Figure 2-2 shows their relative locations in the visible-light range.

Astronomers employ a device called a spectroheliograph to photograph the sun in light at specific wavelengths, corresponding to Fraunhofer bands in the spectrum. These photographs show a sun with many different "faces." Some areas of the sun appear much brighter in calcium light or hydrogen light, for example, than other areas; these differences do not show up in ordinary photographs

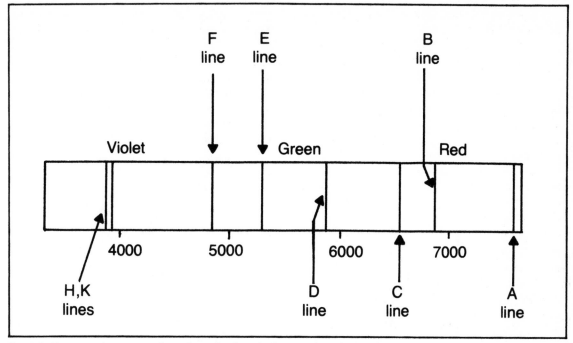

Fig. 2-2. Common elements produce absorption lines in the spectrum of the sun.

of the sun. The reasons for these bright and dark patches are not fully understood, and they present another of the many puzzles of our own parent star, the sun.

BLEMISHES ON THE SUN

You can view the sun through an ordinary telescope by projecting its image, from the eyepiece, onto a screen as shown in Fig. 2-3. This technique is entirely safe for the eyes, and has the advantage of allowing observation by several people at the same time. When looking at the sun's disk, you will often see that it is not a perfectly flawless ball. Sunspots, which appear dark against the rest of the sun, can often be seen with simple equipment such as a small telescope. Sometimes sunspots grow to be thousands of miles across, and can be seen with a simple pinhole device, or even during a sunset.

Ever since sunspots were first seen, astronomers have been somewhat puzzled by them. Magnetic fields have been observed in association with the spots, which almost always appear in groups. Sunspots were used to obtain the first de-

termination of the speed of rotation of the sun. The spots move across the disk from east to west, but they move faster at lower solar latitudes than at higher ones. Thus we know that the sun rotates faster at the equator; the equatorial period is about 25 days, but the period at high latitudes is about 35 days.

All sunspot groups have magnetic fields around them. One pole of this field is near the leading spot in the group, or the one furthest toward the west as we see it. The other pole is near the trailing spot.

Strangely, sunspots do not form at random. A definite pattern is observed; sunspots do not occur very close to the equator or the poles of the sun. Some spots last for only a few days, while others last for weeks or even months. It seems that the sunspots are like giant hurricanes or typhoons of magnetic origin. There is evidence that great violence is associated with sunspot groups.

The number of spots on the sun fluctuates in a regular cycle that has a period of about 11 years. This cycle has been noticed for centuries, but pre-

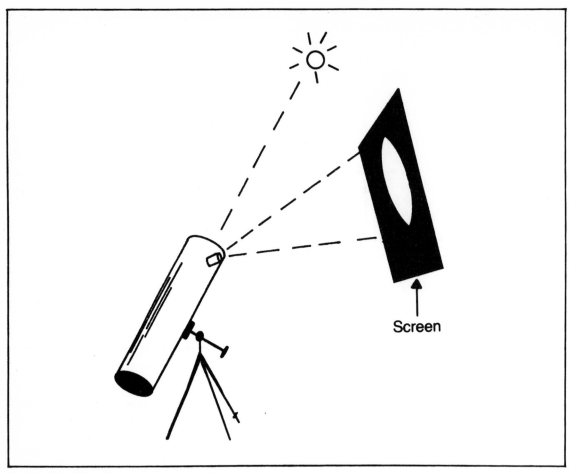

Fig. 2-3. The disk of the sun may be viewed by projecting the image from a telescope onto a screen.

cise plotting of the sunspot activity has been carried out for only the past few decades. Figure 2-4 shows the most recent, and forecast, sunspot maxima and minima. The reason for this cycle is not fully understood, but one odd thing has been seen: With each cycle, the polarity of the sunspot groups reverses! If the leading spot in a northern-hemisphere group has a magnetic north pole during one cycle, it will have a magnetic south pole during the other cycle. The polarity in the southern hemisphere of the sun is always opposite to that in the northern hemisphere. The sun thus seems to act, in a sense, like an electromagnetic oscillator with a period of 22 years. The magnetic field of our own earth has been known, as well, to occasionally reverse its polarity.

Sunspots are not the only imperfections on the visible surface of our parent star. Occasional bright spots have been seen. These are called solar flares. Such eruptions seem to be most common during times of maximum sunspot activity, and they usually take place near large, unstable sunspot groups. The effects of a solar flare are noticed on our planet within an hour or two.

When a solar flare erupts, subatomic particles are thrown off into space. These particles are attracted to the magnetic poles of our earth, and cause interference to the normal ionospheric conditions. The magnetic field of the earth will sometimes fluctuate in intensity following a solar flare; this is known as a geomagnetic storm. Radio communications at the shortwave frequencies are often wiped out for several hours or days; at night, we can see the ionization in the upper atmosphere as a faint glow that takes strange, moving forms. This display is, of course, known as the northern lights to natives of North America and Europe.

At the edge of the visible disk of the sun, where it is easy to see any upward or downward movement of glowing gases in the solar atmosphere, bright red streamers called prominences are sometimes observed. These were first noticed during total eclipses of the sun, when the moon blocked out the bright disk to allow viewing of the dimmer solar atmosphere. Today, astronomers have devices that allow viewing of the sun's atmosphere at any time. Prominences are common; they often rise to heights of thousands of miles above the visible surface of the sun. Occasionally, a large

prominence will be thrust as much as a million miles into space.

As seen from an imaginary space ship in a low orbit around the sun, a prominence would be an awesome display. It would rise to a height of many times the earth's diameter. Figures 2-5 and 2-6 show what we might see from such a vantage point. We would, of course, have to look through dark tinted windows to see anything; otherwise, the extreme brilliance of the solar surface would blind us in seconds. Our ship would pass unhindered through the prominence; the sun's atmosphere is far less dense than that of the earth.

THE BRIGHTNESS OF THE STARS

A few of the ancients may have dreamed that the distant stars were other suns, like our own. But this fact has been known, with certainty, for only

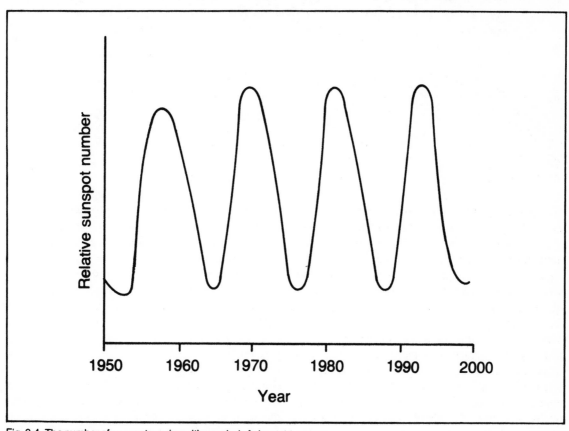

Fig. 2-4. The number of sunspots varies with a period of about 11 years.

Fig. 2-5. From the vantage point of a low orbit about the sun, prominences would appear formidable (courtesy of Big Bear Solar Observatory, California Institute of Technology).

about 100 years. The spectroscope was an important aid in making this discovery. Other stars show signs of the same elements that we see in our sun and on our planet. But one characteristic of the nighttime stars has always been obvious even to the most untrained eye, and must have been a source of bewilderment to ancient astronomers: Some stars are much brighter than others.

Nearly 2,000 years ago, the astronomer Ptolemy catalogued the stars according to their brightness. This was a subjective process. The brightest stars were assigned the classification of magnitude 1, or first magnitude. Dimmer stars were called second magnitude, third magnitude, and so on up to the sixth magnetide. A sixth-magnitude star was just barely visible to the un-

aided eye. Later, when the telescope was invented, the system of Ptolemy was extended to fainter stars, made visible by the larger aperture of the telescope objective lens. In 1856, a formal system for defining star magnitudes was proposed. This has become the basis for our present method of defining star brilliance.

A difference of one magnitude is a change in brightness of 2½ times. Therefore, a first-magnitude star is 2½ times as bright as a second-magnitude star; a second-magnitude star is 2½ times as bright as a third-magnitude star, and so on. One example of a first-magnitude star is Spica, in the constellation Virgo. Magnitudes are defined down to the tenth or even hundredth part. Regulus, in Leo, has a visual magnitude of 1.36. Polaris, the

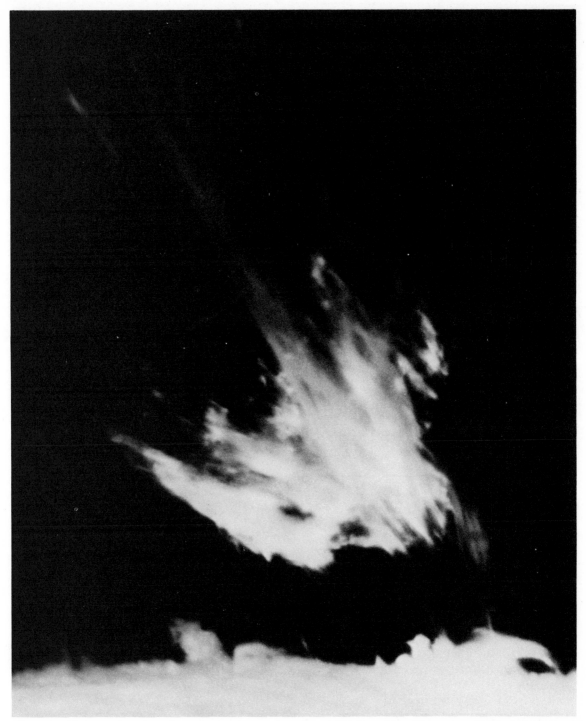

Fig. 2-6. This is another view of a solar prominence (courtesy of Big Bear Solar Observatory, California Institute of Technology).

north-pole star, has a magnitude of about 2. Some stars are brighter than the first magnitude; this is a sort of accident in the definitions of magnitudes. Arcturus, in the constellation Bootes, and Vega, in Lyra, have magnitudes of about 0. The brightest star outside of the solar system is Sirius, in Canis Major, with a magnitude of −1.43.

Of course, it is not hard to guess that the brightness of a star, as we see it from the earth, depends on two things: its actual luminosity, and its distance from us. A very bright star might look much dimmer than a tiny star, if a brighter star is far away and the tiny, dim star is very close by. The observed magnitudes of the stars are called their apparent magnitudes. The actual brightness of a star can only be determined if we know its distance. Astronomers have developed a system of absolute magnitudes, that tells us how bright a star really is. By definition, the absolute magnitude of a star is the apparent magnitude we would give it, if it were exactly 32.6 light years away. Sometimes, then, the absolute magnitude of a star is much brighter (lower) than the apparent magnitude; such is the case for distant stars. For nearby stars, the opposite is true.

One of the brightest known stars is Canopus, in the southern celestial hemisphere. This star can only be seen south of 38 degrees north latitude, or about the location of San Francisco, California. This star has an absolute magnitude of −4.4. Our sun has an absolute magnitude of +4.8; this means that Canopus is almost five thousand times as bright as our parent star! If Canopus were suddenly substituted for the sun, we would all be burned to death within a few minutes or seconds. But because of its distance, Canopus does not appear unusual in the sky. At a distance of 32.6 light years, the sun would be a faint star, visible only away from city lights.

DISTANCES TO THE STARS

The actual distances to the stars remained a mystery until the advent of the telescope. This instrument gave astronomers the ability to measure very small angles in space. When this became feasible, astronomers set out to find the distances to some of the stars. Triangulation was used for this purpose. As the base line for their observation, the greatest possible distance was chosen: the diameter of the earth's yearly orbit around the sun. The longer the base line, the more accurate the distance determination by triangulation. Figure 2-7 shows how the distances to the stars are measured. Patience is a necessary virtue in this kind of project; a six-month wait is required between observations!

In the process of measuring the distance to a star, an astronomer must choose a background of

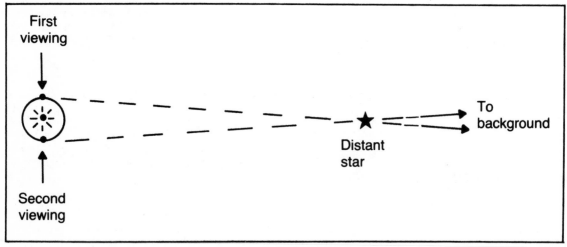

Fig. 2-7. The distances to some stars can be measured according to the parallax that occurs as the earth revolves around our sun.

stars so distant that their relative positions do not appear to change as the earth revolves around the sun. This is fairly easy to do with photographic equipment. A portion of the sky that lies 90 degrees away from the sun can be photographed, and later re-photographed after six months have passed. By choosing the proper times of year to take the photographs, any part of the sky can be evaluated. When the photographs are compared, some stars appear to have changed position with respect to the background majority; these stars, within the range that allows determination of their distances by parallax, are the nearest stars in our galaxy. If a star shifts position by 1 second of arc—1/60 of 1/60 of a degree—then the star is said to be at a distance of 1 parsec. All stars are farther away than 1 parsec (the word parsec is short for parallax second). A parsec is about 3.26 light years, or 19.2 trillion miles. The triangulation method is useful for measuring star distances up to about 100 parsecs, or a parallax of 0.01 second of arc.

The nearest star to our solar system is Proxima Centauri, near the bright double star Alpha Centauri. Proxima Centauri is about 4.3 light years, or 25 trillion miles, away. That is an enormous distance compared to the separation of our planet from the sun. Imagine the orbit of our earth scaled down to the size of a nickel. On that scale, Proxima Centauri would be 1¾ miles away! The sun, in proportion, would be about the size of a grain of sand; the earth would be so small that a microscope would be needed to see it. The limit of our ability to measure star distances by parallax methods—about 300 light years—would be represented by a sphere about 120 miles in radius. The diameter of our Milky Way galaxy on such a scale would be approximately 40,000 miles, or five times the diameter of the real earth. The Andromeda nebula would be 20 times more distant than this, or three times as far away as the real moon. If only the earth's orbit around the sun were just as big as a nickel!

TYPES OF STARS

The first serious attempts to study the spectra of stars revealed dark absorption lines, as we see from our own sun. But not all stars have the same pattern of lines. In the late nineteenth century, at the Harvard Observatory, an astronomer named Annie J. Cannon compiled a catalogue of the spectra of nearly a half million stars. This became known as the *Henry Draper Catalogue*. There are seven main spectral types of stars. They have been given the names O, B, A, F, G, K, and M. More subtle differences in stellar spectra can be noticed; each of these seven classes has been subdivided into ten parts, indicated by a number from 0 to 9. Thus a type A9 star is followed by a type F0 star, and so on; this results in 70 different stellar spectral classifications!

The spectrum of a star tells us, very accurately, the surface temperature. The type O stars are the hottest, and appear blue to the eye. The type M stars are the coolest; they look reddish in color. With a particular alphabetic subdivision, the number 0 is the hottest and 9 is the coolest.

Our sun is a type G2 star. Thus it is somewhat toward the cool end of the spectral continuum. The sun is also a fairly small star, in terms of physical size. A natural question thus arises: Is the absolute brightness of a star related to its surface temperature? The answer is yes. This relationship was found early in the twentieth century by a Danish astronomer named Ejnar Hertzsprung, and an American, Henry Russell. These two scientists independently plotted the absolute magnitudes of nearby stars on a graph, as a function of their spectral classification. They found that, in general, the hotter a star, the brighter it is. This fact is not too surprising. But there are exceptions to this rule.

THE HERTZSPRUNG-RUSSELL DIAGRAM

When the spectral type of a star is graphed along with its absolute magnitude, the star is represented by a single point. Figure 2-8 shows what Hertzsprung and Russell found as they put their graphs together. Such a graph is called, appropriately enough, a Hertzsprung-Russel diagram. On the horizontal axis, the highest temperature is toward the left, and the coolest is toward the right. On the vertical scale, the brightest absolute magnitude is toward the top, and the dimmest is toward the

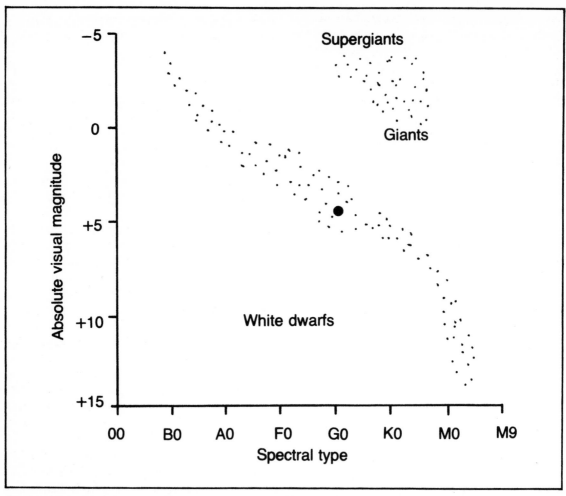

Fig. 2-8. Most stars fall along a characteristic curve in the Hertzsprung-Russel diagram. This curve is called the main sequence. Our sun is a main-sequence star (large dot).

bottom. Our sun, a very average star, is shown by the large dot.

Stars in the upper right part of the Hertzsprung-Russell diagram are shown as red giants Stars in the lower left portion are called white dwarfs. Blue giants are in the upper left-hand corner. The smallest and coolest stars, the red and orange dwarfs, are at the lower right. As stars were plotted on this chart, an interesting phenomenon was noticed: Most stars fall along a somewhat S-shaped curve running diagonally from the upper left to the lower right. This region of the diagram,

containing the majority of stars, is called the main sequence. The red giants, to the right and above the main sequence, and the white dwarfs, below and to the left, are the main exceptions.

When astronomers began investigating the relationship between the location of a star in the galaxy and its position in the Hertzsprung-Russell diagram, some fascinating discoveries were made. Hot, massive stars seem to be concentrated mostly in the flat, disk-shaped part of the Milky Way, in the spiral arms but not in the central regions. The spiral arms happen to be where most of the interstellar dust and gas is found. Could it be that stars are forming from this material? Astronomers believe so. Near the center of the galaxy, there is relatively little interstellar debris, and stars there are frequently found in large congregations. It is thought that stars in the central part of the galaxy are much older than the stars in the spiral arms.

The relationship between star types and their location on the Hertzsprung-Russel diagram has helped in the effort to unravel the mysteries of star formation and evolution. We find much dust and gas in the plane of the galaxy; this is where we should look, then, to see stars in the formative and early stages. Most of the interstellar material is gone from the center of our galaxy, and from regions above and below the plane of the spiral. Here, we should expect to find older stars.

BIRTH OF A STAR

All stars evolve from clouds of gas and dust. If the original material in the universe were perfectly homogeneous—equally dense at every point—perhaps stars would have never formed. But this was, fortunately, not the case! The tenuous wisps of matter in space are more dense in some places than in others. This has been true from the very beginning, since just after the big bang.

Where the clouds of matter were the most dense, the gravitational attraction among the atoms was the greatest. This caused the dense regions to become even more dense, and the sparse regions to get more sparse. A vicious circle ensued, and it happened in many discrete locations. It is still evidently taking place in the spiral arms of our galaxy and other galaxies.

As a cloud of gas and dust contracts, it eventually starts to heat up. The atoms, originally free to move almost without restriction, get cramped for space and start to collide with each other. But the force of gravitation becomes much more powerful than anything the heat can counterbalance. The cloud begins to glow because of the heat. Energy is radiated from a point near its center. Gravitation continues to pull the cloud into a tighter and tighter mass. Eventually, the temperature at the center of this huge congregation of atoms, consisting mainly of hydrogen and helium, gets so high that nuclear fusion begins. The nuclear reaction generates far

more heat and light, as well as other radiant energy, than simple compression. The force of gravitation has met its match! The contraction process slowly grinds to a halt. Figure 2-9 is an artist's conception of what this process looks like on a time-lapse scale. Several hundred thousand years, or perhaps millions of years, go by between the beginning of the contraction and the start of the nuclear-fusion reaction. Massive clouds contract more rapidly than small clouds. Large stars are born more quickly than small ones. Sometimes, there are several centers of contraction; this might, in fact, be the rule rather than the exception. Then, several stars are born in a cluster. The Pleiades, a familiar group of stars called the Seven Sisters (since seven of them are visible to a good pair of unaided eyes), is a cluster of young stars that have formed fairly recently from interstellar debris.

We can follow the metamorphosis of an embryonic star in a Hertzsprung-Russell diagram. A single protostar, as it contracts and becomes hot, is not very luminous until the fusion reaction begins. Therefore, the protostar is initially situated at the lower part of the diagram, corresponding to low brightness. As the protostar contracts to the point where fusion starts up, its position in the diagram moves toward the top. Almost every new star comes to rest on the main sequence. The most massive stars end up at the upper left, and thus become bright and hot. The least massive stars end up at the lower right, and become rather dim and cool. Figure 2-10 shows this process on the Hertzsprung-Russell diagram. Oddly enough, the dim, cool, and least spectacular stars are the longest lived. The big, massive, hot stars generally face short life spans.

Once the fusion process begins and the star begins to glow brightly, the remaining gas and dust that surrounds the star becomes ionized because of ultraviolet radiation. When we look at the Pleiades, for example, through a large telescope, this glowing gas can be seen. Gravitation eventually pulls some of this remaining material into the young star or stars. The rest of the debris is blown away by radiation. The material may then form other stars. The interstellar medium is in a constant state of

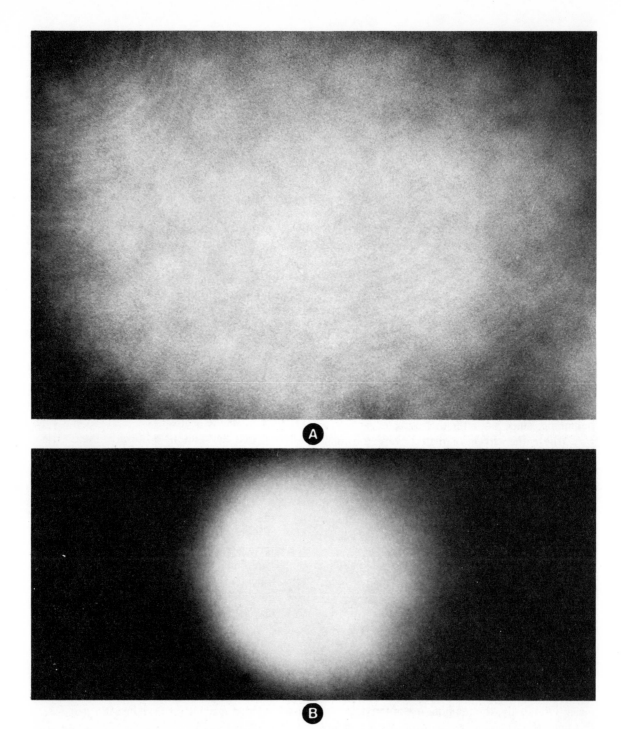

Fig. 2-9. As a cloud of gas and dust collapses, it glows more and more brightly. Finally, nuclear fusion begins. Here, a diffuse cloud is shown in the drawing at A. The cloud collapses to progressively denser states (B and C), until finally the temperature rises to a level sufficient to start fusion (D).

C

D

53

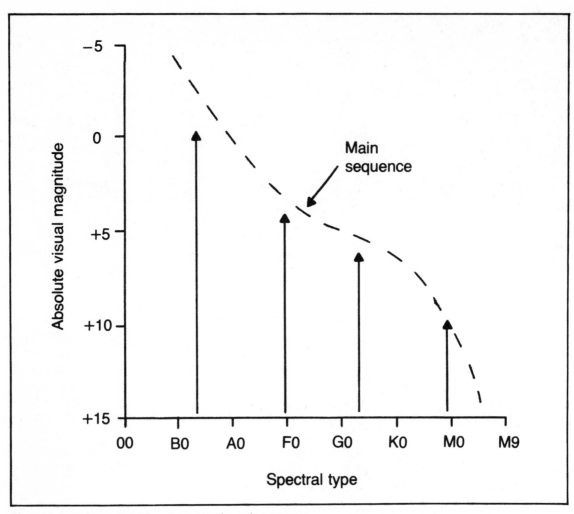

Fig. 2-10. As stars are born, they move onto the main sequence.

turmoil, blown about by radiation and influenced by gravitation, like smoke in a large room full of restless people.

LIFE OF A STAR

Astronomers have discovered that massive stars lead much different lives than small ones. While the smaller stars are not nearly as spectacular as the supergiants, they last for a much longer period of time. Our sun seems to be near the dividing line between the extremely stable red and orange dwarf stars and the comparatively volatile larger stars.

Eventually, most of the hydrogen in the center of a star has been converted to helium by the fusion process. For nuclear reactions to continue, the core temperature must rise. This necessitates contraction of the star. Once the center of the contracting star becomes hot enough, fusion begins among the helium atoms in the center. This creates heavier elements. Our sun has not yet reached this stage.

It is believed that most, if not all, of the complex matter in our universe was built up in the centers of stars now dead. That includes such essentials for life as carbon, nitrogen, and oxygen. It includes the rock and metal of our planet, and the

54

salt in our oceans and in our blood! It includes all the fundamental building blocks of our bodies.

Stars that weigh less than the sun generally evolve slowly and quietly, until they can no longer support nuclear fusion reactions. Then, they cool off and contract into very dense planet-like objects. The larger stars, that have more than about 1.5 times the mass of our sun, await a much shorter lifespan. They often die a violent death; their explosions are probably responsibile for the heavy elements that are seen in interstellar debris. It could well be that our existence is dependent on the explosions of ancient stars! A universe with nothing but small, quiet stars might well be a lifeless cosmos. None of the heavy elements would ever find their way into the interstellar medium of such a universe, to form planets like the earth.

It is the larger stars, then, that have attracted the most attention from astronomers. Some stars are many times as massive as the sun. These stars often eventually swell to extreme size, and their surface temperatures cool. Such stars are called red giants. Some red giants are larger than the diameter of our earth's orbit around the sun! It is possible that our sun, in a few hundred million years, will grow into a red giant of this stature. Red-giant stars meet either of two kinds of doomsday: They either contract and become tiny and dense, or else they explode. In the former case, they end up as white dwarfs. In the latter case, the cataclysm is called a supernova.

ANATOMY OF A STAR

It is doubtful that we shall ever be able to travel inside a star to find out precisely what its interior is like. We may someday venture into interstellar or even intergalactic space, but the interior of a star is incredibly hostile! A science-fiction story by the Russian author Henrik Altov, called "Icarus and Daedalus," presents a fascinating look at a hypothetical trip through the center of the sun. But it will probably always be fiction. Scientists know, nevertheless, some things about the details of stellar interiors. Different types of stars have different internal structures. We also know that the effects of a star extend far beyond its bril-

liant visible surface, in the form of a glowing halo called the corona.

In recent years, electronic computers have been used by astrophysicists to construct what they call model stars. The known physical laws are applied to explain what will happen inside a star having a certain mass and material composition. Stars are considered to be made of many concentric shells of matter, each with a certain temperature, mass, and combination of elements. The whole star is "built" by putting all of these shells together. This can be done in either of two ways: outward from the center toward the surface, or inward from the surface toward the center. Sometimes the two methods yield different model stars.

Even with the knowledge presently available about the behavior of matter and energy, the model stars resulting from these computerized simulations often do not turn out quite the same, in terms of observable properties, as real stars. The way in which energy is transferred from the center to the surface of a star may change at various depths. And this depends on the mass of the star as a whole.

In the depths of a fairly large, massive main-sequence star, radiation and convection are believed to be mutually responsible for the transfer of energy from the core of the star into space. But, near the surface, radiation predominates; near the center, convection is believed to be more powerful, since the material deep within a massive star is more opaque to radiation. This model of a large star is shown in Fig. 2-11.

In a much less massive star, the reverse is believed to be the case. A red-dwarf star is thought to have a convective outer region, with a more transparent inner core. Radiation thus predominates deep within the star, where convection is more dominant in the outer regions. This is shown in Fig. 2-12.

The sun is, as we have seen, an average star in terms of its location on the Hertzsprung-Russell diagram. What is the interior of the sun like? There is one great advantage when the sun is used for the purpose of evaluating star interiors: We can observe the sun in fairly great detail, compared to the other stars in the universe. Astrophysicists have

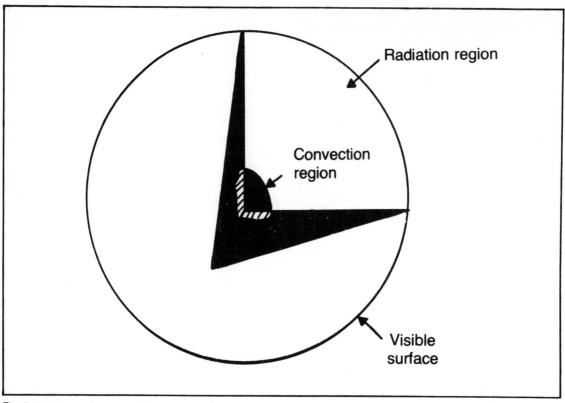

Fig. 2-11. A large star generally has a convective core, surrounded by a region in which energy transfer occurs primarily by radiation.

constructed several different theoretical models of our parent star. Although there is some uncertainty about the details of the interior anatomy of the sun, all the different models tell us that the temperature at the center of our parent star is about 25 million degrees Fahrenheit. At the surface of the sun, we can see from spectroscopic examination that the temperature must be about 10,000 degrees Fahrenheit. Above the visible surface of the sun, the temperature rises again, to about a million degrees Fahrenheit; outward from the immediate vicinity of the star, the temperature steadily drops until it is near absolute zero at a distance of several billion miles. The interior of the sun probably resembles the red dwarf (Fig 2-12) more than the anatomy of a larger main-sequence star (Fig. 2-11). This has led to the belief that the sun's death will be a rather quiet event—quiet, that is, compared to what it could be if the sun were a very large star.

Although it appears unlikely that the sun will become a supernova, we know that it will not continue to shine forever as it does now. In a few hundreds of millions of years, or perhaps after several billion years, there will be a slow but radical change in the sun and its effects on the earth. The change, slow as we would perceive it, is actually destined to be quite rapid on the scale of stellar evolution.

After hundreds of geologic summers and winters, times of warmth and cold on the earth, it is believed that the sun will first expand and grow brighter, and then contract and grow dimmer. Its ultimate end will probably render it only about as big as the earth, and just as cold and black in the emptiness of space. But by then, man will not miss the light and warmth of the sun. If our race still exists, it will have long since found another home, in the planetary system of another younger star.

DEATH OF THE SUN

According to astronomers and astrophysicists, our sun still has plenty of hydrogen fuel left. Our parent star should keep on shining, just as it shines today, for many millions of years to come. But eventually, the supply of hydrogen will begin to run out. This will occur at first at the center of the sun, and then in layers that migrate progressively outward toward the surface. The core of the sun will, at some distant future time, begin to shrink because of gravitation, until the pressure rises sufficiently to start helium fusion reactions. The outer layers of the sun, still containing plenty of hydrogen, will begin to undergo fusion into helium. The increased temperature of the sun, both at the core and in the outer layers, will cause the whole star to swell in size. In the center of the sun, helium atoms will combine to form heavier elements such as carbon and oxygen.

As the sun expands because of increased pressure from deep within, the surface will thin out and cool off. The color of the brilliant photosphere of the sun, now considered yellow, will deepen to yellow-orange, then orange, and finally to a reddish orange hue. The apparent size of the sun's disk in the sky will get larger and larger. Science-fiction writer H.G. Wells foresaw this strange expansion of the sun in his story "The Time Machine," even before scientists knew, or could predict, that the sun would change in this way. The two nearest planets to the sun, Mercury and Venus, will probably be swallowed up as the parent star swells. The earth, too, may be engulfed. The sun will become a true red giant in its old age.

These changes will happen over a period of many human lifetimes. But in a frame of the sun's life span, the metamorphosis will be rapid. The first noticeable effect on our planet, as the sun begins to

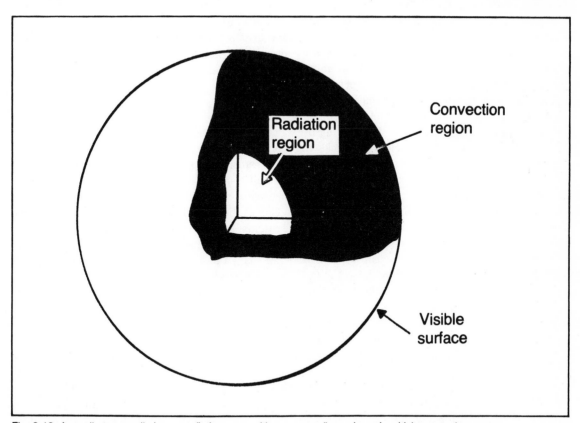

Fig. 2-12. A small star usually has a radiating core, with a surrounding volume in which convection occurs.

expand, will be a slight increase in the mean temperature. This will be accompanied by increased rainfall over much of the planet, and perhaps desertification in some regions. The ocean level will rise as the polar ice caps melt. Low-lying land areas will be come part of the sea floor. The frequency and intensity of storm systems will probably increase. The weather may become violent indeed, and more hostile than anything we have ever known. Hurricanes of incredible ferocity will probably roam the oceans, slamming into the continents to cause indescribable destruction. Great electrical storms will rage over the mountains and prairies. Tornadoes will wander about the countryside. But blizzards will be unknown!

As the temperature on our planet inexorably rises, the atmosphere will begin to thin out as the heat and radiation of the sun blows it off into space. The water in our lakes, rivers, and oceans will boil, and escape with the air into the interplanetary void. Eventually, all of the air and water, those elements so important to the sustenance of life, will be gone from the earth, and our planet will resemble the present surface of Mercury. The sun will fill more and more of the sky; instead of ½ degree of arc, the brilliant disk may get to be so large that it fills the whole sky. Even at night, the glow of the brilliant sun will be dimly visible because of scattering by the many billions of dust particles and rocks in interplanetary space. The moon, if it still exists, will have a bizarre bright red-orange glow. The distant planets—Mars, Jupiter, and Saturn—will become brighter and redder. Even Uranus, scarcely visible now to the unaided eye, and Neptune, invisible now without a telescope, may seem bright in the sky. But, of course, no man, woman, or child will stand on the earth to view these sights.

Finally, all of the helium fuel at the core of the sun will be used up. The sun will once again begin to collapse at the center. The rest of the sun will follow, and the red giant will shrink back to a more reasonable size. The internal temperature, as well as the surface temperature, will rise. Further sporadic fusion cycles may occur, causing the sun to alternately expand and contract over thousands or millions of years. Ultimately, all possible nuclear reactions will have been exhausted, and there will be no more fuel for the sun. The parent star will then shrink in size until it is a dense, dim, white ball called a white dwarf. The radiation from this little star will lessen, gradually, until the sun attains its final identity: a black dwarf.

We may argue that it is rather silly, after all, to begin planning our escape from the earth for that distant time when the sun reaches the red-giant phase! There are more pressing troubles at hand right now. But it is nevertheless interesting to speculate about where we might go when this irreversible change finally comes over our sun. It is fun to think about what our descendents might do to survive that cataclysm, assuming our human race can last that long. Perhaps the planets Jupiter or Saturn will evolve into earthlike places as the sun gets too hot for our present abode. In a billion years or so, man may have perfected interstellar or intergalactic travel. It might be no problem to move on. Perhaps at the time our sun begins to grow hotter and brighter here, we will have a list of several thousand possible alternative dwelling places in the galaxy. Maybe some of these places will already be colonized!

VARIABLE STARS

Although our parent star will eventually become sick and die, just as human parents do, the sun is a remarkably healthy and well-tempered star. Not all stars are so peaceful. There are many pulsating, or variable, stars in our galaxy and in other galaxies. Some of these fluctuating stars are binary systems that eclipse each other. But many, or most, are stars that actually get brighter and dimmer. This often happens at an extremely regular and predictable rate. The Cepheid variables are especially interesting, because of the clockwork regularity of their changes in brilliance.

In the case of the Cepheid variable stars, so named because one of the first discovered and most well known of them is the constellation Cepheus, astronomers have found that the period and the luminosity are correlated to an astonishing precision. The longer the period, the brighter the star is, on the average. As these variable stars get brighter and dimmer, they also get hotter and cooler, and they expand and contract in size. We know that the temperatures of the Cepheid variables change, because of fluctuations in the spectral characteristics. We know that the Cepheids change in size, even though they are all too far away to appear as disks in the most powerful telescopes. The Doppler effect has been observed in their spectral lines, as the star surfaces first approach us (as they expand) and then recede from us (as they contract).

The periods of the Cepheids vary. Some of the stars have a cycle that requires less than a single day to complete. Others have periods of several weeks. Some of the Cepheids change brightness to a large extent, and others hardly change at all. Polaris, the star that we use to mark the position of the north celestial pole, is a Cepheid variable star, but its brilliance fluctuates only a little, so that it is not noticeable to a casual observer.

Cepheid variable stars are always massive yellow stars, known as yellow supergiants. There are other kinds of variable stars; the RR Lyrae type star is an example. This name is derived from the location, in the constellation Lyra, of one of the best known of these stars. None of the RR Lyrae variables are visible without a telescope. The RR Lyrae stars differ from the Cepheids in that they are blue, rather than yellow, and are much less luminous. But, like the Cepheid variables, the RR Lyrae stars have a regular, constant cycle of pulsation.

The relationship between the brightness and the periods of the Cepheids has made these stars exceptionally useful for the purpose of measuring great distances in space. The parallax method of gauging interstellar distance is useful only up to about 300 light years. Beyond that distance—only about 0.3 percent of the diameter of our galaxy—parallax angles are too small to be measured. Once the brightness-to-luminosity relation for the Cepheids became well-known, astronomers used these "beacon stars" to gauge vast distances, not only within our own galaxy, but among other galaxies.

Why do some stars pulsate, while others do not? Apparently, the Cepheids and RR Lyrae stars oscillate because of some instability that has existed since their very birth. As a variable star contracts in size, it gets hotter. The heat produces an outward force; eventually this force becomes so great that the star begins to expand. Then it cools, and the heat pressure lessens. Finally, gravity takes over again. Like an oscillating spring, the Cepheids and RR Lyrae stars are in a sort of perpetual condition of regulated instability! But why do some stars develop into variables, while others do not? It might have something to do with the motion of the gas and dust prior to the formation of the star, but that is only a guess.

Cepheids and RR Lyrae systems are not the only kinds of variables. They are among the most interesting, because of the relationship between the oscillation period and the absolute brightness. Some variable stars have much longer periods, and oscillate for other reasons. Mira type stars get their name from a star in the constellation Cetus. Mira is a red supergiant. Its radius is greater than the distance from the sun to Mars. This star varies in brightness over a period of about a year, and fluctuates between about the 8th and the 3rd visual magnitudes. At its maximum brightness, it is 100 times more luminous than at minimum.

Astronomers believe that the Mira type variables are not pulsating in size, like the Cepheids or the RR Lyrae stars. Rather, it is thought that alternating swells of hotter and cooler gases migrate outward through the star, from the center to the surface. When the Mira type star is at its dimmest, the surface may be veiled by somewhat opaque clouds. The Mira type variables have fairly regular periods, though; this indicates that their fluctuations are caused by something other than sheer coincidence. But the details are not well known.

Still other kinds of variable stars have been observed, with irregular changes in brilliance, both in terms of the amount of luminosity change, and with respect to the period of the fluctuation. Betelgeuse, the familiar red star in the constellation Orion, is a variable star with an irregular period. We do not normally notice its changes in magnitude, since they are not very great. The irregular variable stars are often monstrous red giants or supergiants, like Betelgeuse. During periods of greater brilliance, such stars throw off clouds of diffuse gas billions of miles into space.

FROM RED GIANT TO WHITE DWARF

The range of sizes and magnitudes of stars is not really surprising when we consider that, statistically, some clouds of interstellar material are bound to be much larger than others. Stars form from these clouds; the variety of sizes and temperatures can be expected to cover a great range.

All stars evolve. The larger stars have shorter lives than the smaller stars. But as a star becomes old, its disposition can change radically. Our own sun will probably swell to the size of the earth's present orbit, once its stable hydrogen-fusion reaction begins to fall apart. After this red-giant phase, and subsequent variations in luminosity and size, our sun will shrink to a ball no larger than our planet, and its light will fade away. In the end, the sun will be completely dark. It will also be unbelievably dense.

In general, the larger a star is to begin with, the shorter its life expectancy. The blue supergiants face a life destined to last only a fraction as long as the life of our sun. The red dwarf stars, at the opposite corner of the Hertzsprung-Russell diagram, may still be shining in their patient, unspectacular way, long after our parent star has gone cold.

Larger stars sometimes explode, throwing off much of their matter in a burst of light and heat. Other stars, in the medium-size category, swell to become red giants. A red giant is tenuous; much of the mass is concentrated into a tiny core surrounded by a large diffuse envelope of gas. Computerized models of red-giant stars show a small, hot core surrounded by a large convection region extending to the visible surface.

While red giants are found above the main sequence on the Hertzsprung-Russell diagram, indicating that they are more luminous than average, there is another substantial group of stars below the main sequence, with much less than average brilliance. These are the white dwarfs. Most of the white dwarfs are very small indeed, compared to main-sequence stars. Some are smaller than the

earth. White dwarfs are far dimmer than our sun, which is not a particularly bright star itself. Because of their faintness, white dwarfs are hard to see, even with a large telescope, unless they are fairly near us. For this reason, we are bound to underestimate the number of such stars. It is believed that about one of every ten stars in our galaxy is a white dwarf.

The structure of the white dwarf is much different than that of other stars. The primary distinction is the density of the white dwarf: A cubic inch of the substance would weigh hundreds, thousands, or perhaps millions of tons on the earth. Imagine what would happen to a piece of white-dwarf matter if it were placed on your kitchen table! No doubt, it would fall right through the table, the floor, and all floors below you; it would probably continue on almost unimpeded to the center of the earth. Then it would oscillate back and forth around the gravitational center of our planet, until friction brought it to a stop. All familiar objects are composed of atoms, and the atoms in turn are made up of protons, neutrons, and electrons. But by far the greatest volume in any normal piece of matter is empty space. The protons, neutrons, and electrons are tiny; on their scale, the nucleus of an atom is far away from the orbiting electrons. But in a white dwarf, there is much less space between the nuclei and the electrons. Still, white dwarfs are not the most dense known objects! Further compression is possible!

Because of the tremendous density of the matter in a white dwarf, it would seem reasonable to suppose that these stars must be solid, rather than gaseous. But such is not the case; white dwarfs are actually a form of ionized gas called a degenerate gas. The electrons, rather than orbiting the same atomic nucleus all the time, have been stripped away and are free to move all about.

White dwarf stars radiate largely by heat conduction through the dense material. Some radiation also occurs, since the degenerate gas is fairly transparent to heat and light. The surface of a white dwarf star is covered by a thin layer of gas, perhaps pulled from space by the gravitational field of the star. In this atmosphere, some hydrogen fusion may occur, and this could be responsible for much of the heat and light that comes from the star. Like a dying ember in a fire, the white dwarf gradually loses its stored energy, and cools off.

A red giant developes because of an increase in the inside temperature of a star. But when all the nuclear fusion reactions have been exhausted, and no heavier elements can be formed, the interior temperature falls. Finally, the inexorable force of gravity, which has been waiting for billions of years, seizes its opportunity to crush the star into a white dwarf.

FROM WHITE TO BLACK

Like a glowing cinder in an extinguished fire, a white dwarf continues to radiate for some time. But the amount of residual energy is finite, and cannot last forever. Gradually, the white dwarf becomes dimmer, until it is a dark, cold planetoid of enormous mass and density. Its size, however, is quite small.

The possibility of visiting and actually landing on a black dwarf star is eliminated by the fact that the gravitation would be more intense, at its surface, than anything we could withstand. Our bodies would be crushed. If we want to get a close look at the surface of a black dwarf, we will have to do it with a telescope. From an orbit perhaps a hundred thousand miles above the surface, we might see a few dim traces of light—feeble reminders of the fact that this object was once a star (Fig. 2-13).

If we ever become capable of interstellar travel and actually have the opportunity to observe a black dwarf at close range, we will probably be very disappointed. There would be no hills or mountains because of the tremendous force of gravitation. There might be an atmosphere, but that would matter little since we could never land and hope to survive.

Our own sun is destined to become a black dwarf in a few billion years. The outer planets will faithfully continue to orbit the sun after its death; the mass of our parent star will still be great, and its gravitational field powerful. The inner planets—Mercury, Venus, Earth, and Mars—will still orbit the sun, too, if they have not been vaporized during the red-giant stage. For thousands of years, philosophers have debated whether the earth will end in fire or in ice. Interestingly, science assures us that it *will* be one or the other, but it has still not been completely resolved. Perhaps it will be both: The red giant will surely scorch our planet beyond recognition, and if the earth's rocks can survive this, they will later freeze in the dark!

The larger a star is to begin with, the more dense the final black dwarf will be. A very massive star will form a black dwarf so dense that the atoms will be radically altered. When the force of gravitation, with consequent pressure, gets intense enough, the electrons in a substance are driven into the protons of the nuclei. When an electron combines with a proton, the result is a neutron. It is believed that some black dwarfs are actually huge balls of neutrons, and these objects have been called neutron stars. Neutrons can be packed so tightly together by gravitation that there is almost no space left between them. In fact, with immense pressure, all the space might disappear, in much the same way as a mass of warm chocolates under pressure will combine into a single, huge blob.

The fate of a massive neutron star can be strange indeed. Science fiction has surely met its match in the gravitational and electromagnetic environment of such a dense piece of matter. Gravity bends space, slows down time, and can overpower all other physical properties of the universe, provided a chunk of matter becomes sufficiently concentrated. In Chapter 4, we will examine the strange effects of gravity out of control.

STAR EXPLOSIONS

The nighttime sky is generally thought to be a constant, except of course for the phases of the moon and the variability of cloud cover. Aside from comets, or eclipses of the sun or moon, most spectacular events in the heavens are seen when stars explode. Without a telescope, we cannot hope to see such an event very often, although some dramatic episodes have been recorded in the past few centuries. Star explosions are called novae or supernovae, depending upon how bright they become. A nova may become as luminous as absolute

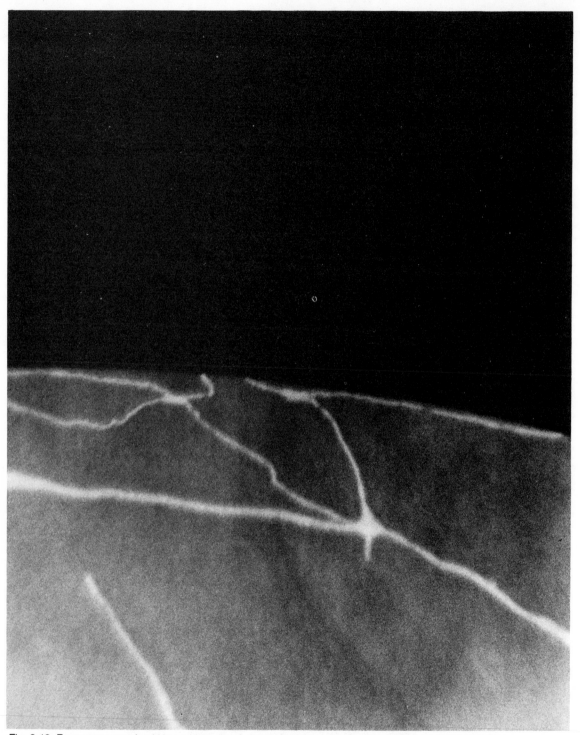

Fig. 2-13. From a spacecraft orbiting a black dwarf, the surface appears dim but smooth, as in this drawing.

magnitude -7, which is perhaps ten times as bright as the brightest supergiants. A supernova can attain a brilliance of absolute magnitude -18 or -19. This is billions of times as bright as our sun.

In the sixteenth century, a supernova was seen in the constellation Cassiopeia, and it attained a brightness comparable to that of Venus. Other supernovae, seen in the eleventh, seventeenth, and early twentieth centuries, have become about as bright as the planet Jupiter. Supernovae have been called "guest stars," and in ancient times they were widely feared. Changes in the absolute heavens, their causes unknown, were believed to portend evil!

Most star flare-ups are too far away to be seen without the aid of a telescope. This is simply a matter of chance. We notice only a very small part of our galaxy when we look at the sky with unaided eyes. Even on the darkest and clearest night, well away from the lights of the city, we can see only perhaps 100 light years. The galaxy contains much gas and dust, and this obscures our view of its stars.

Occasionally, star explosions are seen in other galaxies. With the telescope, this kind of event is fairly common, but one supernova in the Andromeda nebula actually became visible to the naked eye. At a distance of two million light years, this supernova was about 1/6 as bright as all of the rest of the stars in its galaxy put together!

Supernovae were studied in detail, probably for the first time, by an astronomer named Zwicky. He and his colleagues began their work in 1933. They observed many different galaxies, looking particular for supernova displays. They were found, in fact fairly often, although it was surmised that a supernova occured in a particular galaxy only once in every 360 years. By the middle 1970s, astronomers had found over 450 supernovae, and their light variations and spectral characteristics were fairly well known. The light output of a supernova increases with great rapidity over a period of just a few days (Fig. 2-14). The maximum brilliance is reached within two to four weeks, and the most luminous period lasts for about a month.

Astronomers have found that supernovae occur in two distinctly different ways, and thus they have categorized them as type I and type II supernovae. The type I supernova is most prevalent among older stars. The type II supernova is more spectacular, and occurs among massive and relatively young stars. This difference in supernova types is known because of the neighborhoods in which the explosions take place. Type I supernovae occur in the central regions of spiral galaxies, and in elliptical and irregular galaxies, where there is very little interstellar gas and dust. Most of the stars in these regions are old. The type II supernovae are found mainly in the arms of spiral galaxies, where interstellar gas and dust is abundant. The stars in such areas are rather young.

There are various theories that have been developed to explain why stars explode. Supernovae always occur during the later stages of the evolution of a star. Some stars age more rapidly than others; these stars are prone to have type II explosions. Some stars age slowly, and these are more likely to undergo type I supernova events. But the exact reasons why stars explode are not altogether known.

Whatever the initial instability in a newly forming star—whatever the genetic flaw—that ultimately makes it explode, it is a good thing that stars explode once in a while. Exploding stars throw off the heavier elements from which life evolves. Only in stars do temperatures get high enough for fusion reactions to create oxygen, carbon, nitrogen, and the other elements necessary for life as we know it. Only when a star explodes do these atoms get freed to form planets, some with atmospheres containing reactive elements, some with rocks having an abundance of rich mineral deposits, some with water lakes and oceans, and some with living organisms.

After a star explodes, it throws a cloud of gas and dust, heated to extreme temperature, violently outward. A classic example of such a gas cloud, which has aroused the curiosity of astronomers ever since it was first noticed, is the Crab Nebula in the constellation of Taurus the Bull. Figure 2-15 is a photograph of the Crab Nebula.

The Crab Nebula has been identified as a source of strong radio noise and X rays; it has a

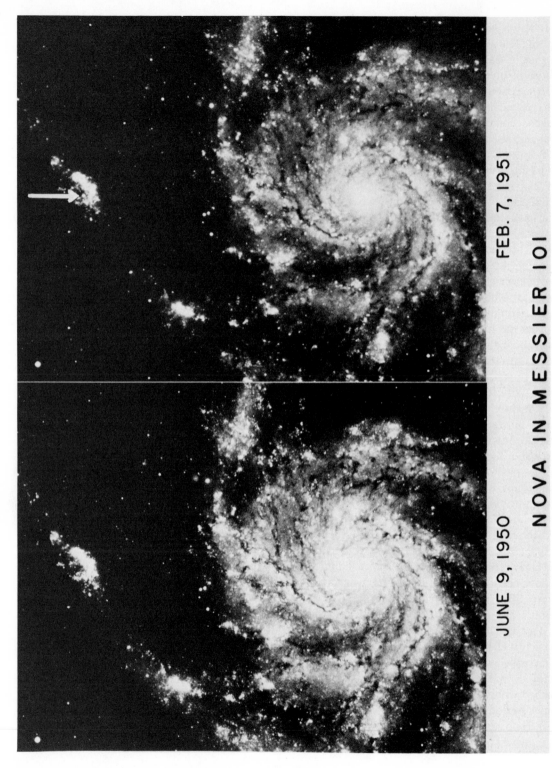

JUNE 9, 1950   **N O V A   I N   M E S S I E R   101**   FEB. 7, 1951

Fig. 2-14. A supernova explosion can be seen from a great distance. At A, the galaxy Messier 101 before a star explodes. At B, a supernova in the same galaxy, shown by the arrow (courtesy of Palomar Observatory, California Institute of Technology).

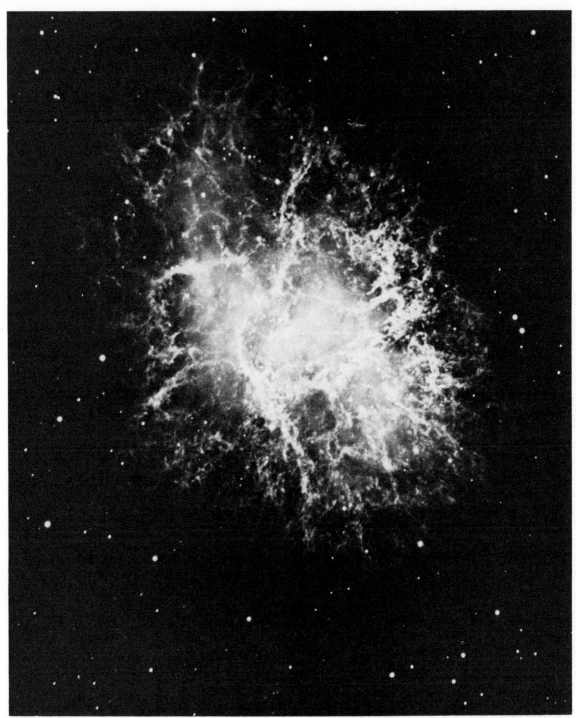

Fig. 2-15. The Crab Nebula in Tauras is believed to be the remains of a star that exploded in the year 1054 A.D. (courtesy of Palomar Observatory, California Institute of Technology).

magnetic field and an intricate structure that has made it an object of study occupying the full-time attention of some scientists. The Crab Nebula has a pulsar at its center; we will look at pulsars in more detail in the next chapter. The Crab Nebula is expanding, as determined by examination with the spectroscope. Its glowing filaments of gas and dust are flying outward from the center at about 950 miles per second. Calculating backward, astronomers have found that this nebula must have begun its outward movement around the year 1100 A.D. In 1054 A.D., a supernova was seen in the constellation Taurus the Bull. The occurrence of the "guest star" is well documented in historical literature. It is believed that the Crab Nebula is the result of that stellar explosion.

Other expanding nebulae have been observed in the locations of ancient "guest stars." The rates of expansion of these nebulae have been determined, and the times estimated for their origination. The calculations yield results that agree with historical records. The expanding nebulae are apparently the entrails of supernovae.

IF THE SUN WENT SUPERNOVA

The sun will probably never explode in the sudden, dramatic temper tantrum of a supernova. Even if the sun does explode, the event will take place in the later stages of the life of our star, most likely billions of years from now. But try to imagine what it would be like on the earth if the sun, our parent star on which we depend for our very existence, were to go supernova during your lifetime!

Supernovae can reach a brilliance of 10 billion times the luminosity of our sun. Again, we flippantly speak of huge numbers and fantastic ratios, so immense that our appreciation of them is blocked because they are incomprehensible. A supernova of the sun, if typical in terms of the rate of luminosity change, would incinerate our planet before it became close to its maximum strength. Even within a few hours, the change would be noticed.

The supernova explosion might begin around five o'clock on a summer afternoon. The first sign that the sun was in a state of flux would be observed on radio-communications circuits. Shortwave

broadcasts would suddenly fade out, because of absorption in the ionosphere. Before long, all but the strongest signals would be obscured by electromagnetic noise, which would sound very much like the static that accompanies a heavy thunderstorm. Even telephone circuits would be affected because of the intense geomagnetic activity. The density of the ionosphere would increase to hundreds of times its normal levels, and on the dark side of the earth, the aurora would be seen.

The brightness of the sun would seem to increase only by an imperceptible amount at first. We see light in a logarithmic way; even a doubling of the sun's visible output would be hard to notice. Beachgoers would find out about the ultraviolet light change, in a painful way, however—their sunburns would be severe by evening!

As the sun set, the twilight would persist for an unnaturally long while. The moon would appear strangely bright, especially if it were near the full phase. The northern lights would put on a spectacular show, unlike ever before (Fig. 2-16). As the night went on, the display would get more and more bizarre; perhaps it would obscure all but the brightest stars, and would be seen as far south as the tropics (and as far north in the southern hemisphere). The moon would seem to grow brighter; it would set in a fiery glow. The first signs of dawn would begin much earlier than usual.

When the sun rose on the second day of the supernova, a flood of heat and blinding light would put the temperature over 100 degrees Fahrenheit by mid-morning. By early afternoon, the heat would become so great that fires would start, and the power would fail as the demand for electricity increased without limit. Fires would burn out of control, and great thunderstorms would rage across the landscape. Killer winds would wreak incredible destruction as the atmosphere of our planet began to escape into space. People might survive that second day to witness an unbelievably brilliant sunset, and a night that never grew dark. But by the third day, all life would end on the earth. The landscape would be scorched. Within a few more days, the entire planet would probably be vaporized by the million-degree heat.

Fig. 2-16. If the sun went supernova during the night, we would be spared a fiery death for a few hours, but not before a spectacular auroral display presented itself to us (U.S. Naval Observatory photograph).

67

We may take comfort in the assurance that our sun will almost certainly never do this to us. If, in several hundred million years, the sun misbehaves in any way, our scientists will probably be able to forewarn us of the event. We will have plenty of time to escape to our colonies on other planets, which orbit other, more stable stars!

STARS WITH PLANETS

One question that remained unanswered, until the advent of sophisticated telescopes, was whether our planetary system is a common phenomenon, or whether the planets formed by some kind of freak accident. The answer to this question dictates the chances of our ever coming across another intelligent civilization; for beings such as ourselves can exist only on planets. It is doubtful that intelligent beings live on any of the other planets in our solar system. None of them appear suitable for life to have evolved to any degree of sophistication. We should hope that planets are common things in this and other galaxies, and that many or most stars have planets.

One of the earliest theories to explain the formation of the planets was the tidal theory. According to this theory, a star passed very close to the sun a few billion years ago. The star came so close that gravitation actually pulled matter from one or both stars, and scattered some of the material into orbits around the stars. As the matter cooled, it condensed into planets. Vortices in the clouds of gas and dust resulted in the rotation of the planets on their own axes, and in the revolution of the planets around the stars.

If this is the manner in which planets usually form, we should not expect to find very many planets in the universe. The chances of a near collision between two stars are miniscule. Stars are tiny compared to the space that separates them. Stars in space are distributed like grains of sand separated by miles! How likely is it that two grains of sand would collide if they were at such a distance? There is, however, a more compelling reason to doubt the tidal theory for the formation of planets. We observe all of the planets revolving around the sun in almost perfectly circular orbits. They are all in nearly the same geometric plane. They all revolve in the same direction around the sun. This sort of orderly arrangement of the planets would probably not result from an encounter such as the tidal theory suggests. Rather, if this is how the planets were created, we would expect that their orbits would be greatly elongated, in different places, and perhaps even in different directions around the parent star.

A more commonly accepted theory of planetary formation suggests that the planets condensed from the same rotating cloud of gas and dust that gave rise to our sun itself. By far the greatest proportion of the mass of this cloud was concentrated, by gravitation, at the center, forming the sun. But some of the cloud remained a considerable distance away from the center, and arranged itself in a disk-shaped region. As the sun condensed into a fairly dense ball, much of its angular momentum was transferred to the material in the cloud. This caused smaller balls of matter to condense in the cloud. The planets were the balls of matter that condensed in this way; they did not have the necessary mass to form stars. Figure 2-17 shows this model.

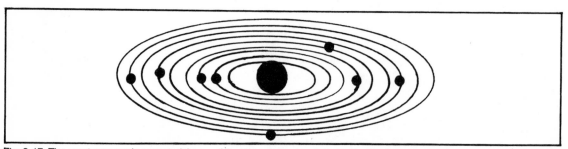

Fig. 2-17. The most commonly accepted theory of the birth of the planets holds that they condensed from a rotating disk of gas and dust at the time the sun was born.

The rotating-cloud theory explains why all of the planets revolve in the same direction around the sun. If the planets condensed from a rotating, disk-shaped cloud of material, they would have to orbit in a uniform direction. The theory also explains, since the cloud was disk-shaped, why the planets all revolve in nearly the same plane. It also explains how the orbits of all the planets can be fairly close to perfect circles. Only Pluto, the farthest planet from the sun, deviates significantly from the circular orbit. Perhaps Pluto was originally an interstellar wanderer, and was captured coincidentally by the gravitation of the solar system. Or perhaps the irregularity of the orbit of Pluto results from the fact that it was toward the outer edge of the disk-shaped cloud of gas and dust from which the solar system formed; greater irregularities would be expected near the periphery than near the center.

If the rotating-cloud theory of planet formation is right—and it is the more plausible of these two theories—then we would expect to see planetary systems fairly often. Many stars should have planets. Chances are very good that many of the stars formed from rotating clouds of gas and dust, and that they could have planets orbiting them. But how do we check for the presence of planets in orbit about distant stars, when even the closest stars are so far away that they appear as dimensionless points in even the most powerful astronomical telescope? How do we go looking for an amoeba-sized object at a distance of five, ten, or more miles? In the scale of interstellar space, it is just about as hard to find a planet in orbit around another star as it is to find a bacterium in a cumulonimbus cloud!

Fortunately, there is a way to detect the influence of a large planet on its parent star, even from quite a great distance. A massive planet, comparable in size to Jupiter, has a very definite effect on the motion of its parent star through the heavens. The geometric center of the orbit of the planet is not exactly at the center of the star; the planet pulls on the star a little bit as it goes around. Therefore, the star should appear to move, if observed from a distance. We would notice the effect of the gravitation of the planet, over a period of time, as a slight wobble or oscillation in the motion of the star. This effect has been seen in the motion of a fairly nearby star, called Barnard's Star.

Barnard's star is quite small, only about 1/6 the mass of the sun. A planet would therefore produce a much greater effect on its motion than it would on a large star, or even on our own sun. Barnard's star is also quite nearby, as stars go. It is only about 5 light years from the solar system. We are able to detect, therefore, very small changes in the movement of this star as it travels through space. The motion of the star does change, and in just the way that we would expect if a planet were in orbit around it. It is belived that there may be two or more planets orbiting Barnard's star, each having a little over half the mass of Jupiter. Of course, there may be smaller planets in the system, too, but their effects are too small to be noticed.

The results with Barnard's star are encouraging, and certainly the search for planets will continue as techniques for detecting stellar motions become more refined. Then, we will be able to observe possible oscillations in the paths of more distant stars.

MULTIPLE-STAR SYSTEMS

It seems necessary that a rotating cloud of gas and dust must condense into bits of debris in orbit around a central mass, in order for the angular momentum to be "shed" so that the central mass does not fly apart because of its own centrifugal force. The bits of orbiting matter can vary greatly in size, from small objects like the asteroids to massive planets like Jupiter. Perhaps Jupiter, if it had been a little bigger, would have become a red dwarf star. Some of the radiation coming from this planet, especially at radio frequencies, is more than reflected solar energy: It is actually generated by Jupiter. Many astronomers think that this planet did, in fact, almost make it to star status. Figure 2-18 shows the planet Jupiter, shrouded by the hydrogen gas that provide the fuel for nuclear fusion in stars.

There is no doubt that double stars, sometimes also called binary stars, are quite common in our galaxy. We can observe them! When we look through a telescope and see two stars that appear to

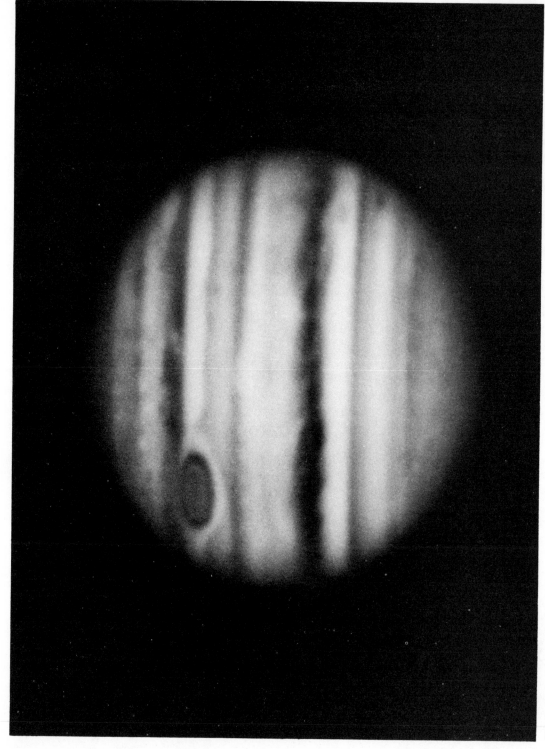

Fig. 2-18. Jupiter, swathed in hydrogen-gas clouds, almost became a star according to some astrophysicists (courtesy of Palomar Observatory, California Institute of Technology).

70

be next to each other, it is possible that they might just be in line by accident. But a binary star system gives itself away over a period of time. Photographs of such a double star system show that the stars are in orbit around each other. We see some binary stars from edge-on, and some at other angles. Figure 2-19 shows what we see at various different slants. It appears that most type G stars, like the sun, become members of double or multiple systems, and that relatively few—about 10 to 20 percent—evolve singly.

The variety of possible binary combinations, in terms of star types and sizes, is just about as great as statistical chances allow. Some binary stars have a very large and massive central member, and a much smaller, dimmer star in orbit around it. Our own solar system would have become a binary system of this kind, if Jupiter had been a little larger and formed a red dwarf. The rest of the planets could still have attained their present orbits, with perhaps a little more eccentricity in the orbits of Mars and Saturn, the planets nearest Jupiter.

Some binary systems evolve into stars that are fairly close to the same size. In such cases, the orbits of planets in the system would be greatly distorted by the gravitational effects of first one star, and then the other; some of the planets might even orbit both stars alternately in a figure- 8 pattern! The environment on such a planet would certainly be variable.

Some binary stars form so close together that they are almost in physical contact as they fly around each other in a fast mutual orbit. In fact, it is believed to be a rather common thing for the members of a binary pair to share their material. If the stars are extremely close, the speed of their orbits causes them to become oblong in shape, and some of the matter may actually be thrown off into space. The evolution process of the two stars in a close binary system proceeds in a mutually dependent fashion. Gas exchanged between the stars may make the initially smaller star more massive than its companion after a few hundred million years.

One particularly interesting kind of binary star system is the eclipsing binary. Such a system is seen edgewise, and has a large, dim star (such as a

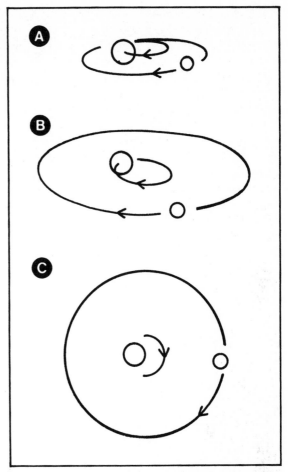

Fig. 2-19. We might see a binary star from almost edgewise (A), from a slant (B), or from the broadside (C).

red giant) in mutual orbit with a smaller, but much more luminous, companion. As the two stars revolve, they alternately eclipse each other. First, the brighter, smaller star passes in front of the dimmer, larger member. This results in a slight drop in the observed luminosity of the system, since the part of the large star that is eclipsed by the smaller star is obscured from view. The drop in brilliance in this case is very small, and may be hardly noticeable. Then the larger star passes in front of the smaller star, totally eclipsing the brighter member of the system. In this case, a precipitous drop occurs in the brightness of the system as we see it. Such a variable star is not the result of actual changes in the brightness of the stars themselves. The eclipsing binary shows an extremely regular rate of change in brightness, and the shape of the luminosity-versus-time curve is characteristic of it (Fig. 2-20).

Some star systems contain more than two members in mutual orbits. In fact, it is believed that four, five, or more stars sometimes form in the same system. Such a star group would probably not be suitable for the evolution of life on any of the planets entangled in their gravitational tug-of-war. The temperature variations on such hapless planets would be not only enormous, but irregular.

The search for life elsewhere in the universe is a major motivating factor behind the pursuit of cosmic knowledge. We will have more to say about the possibilities of, and the chances of finding, intelligent extraterrestrial life in the last chapter.

While perhaps 80 to 90 percent of all stars are members of multiple systems, and therefore do not represent very good places for the evolution of life forms, that still leaves 10 to 20 percent of the stars as candidates for planetary systems hospitable to life. Maybe only one such star in a million has a planet like the earth, or with an environment suitable for the evolution of some other form of life. In our galaxy alone, then, there would be 20,000 to 40,000 different places to look! That is a pessimistic estimate. There may be many more planets with environments good for the development of intelligent life.

STAR CLUSTERS

Some star systems, while not actually in mutual orbits, nevertheless contain hundreds or even thousands of individiaul constituent stars, all

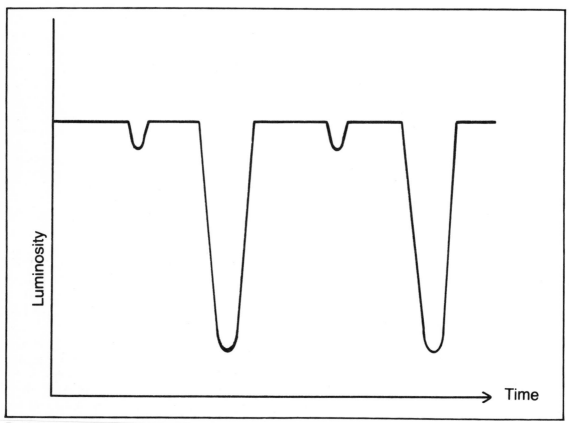

Fig. 2-20. An eclipsing binary has a distinctive brightness-versus-time curve.

under a common gravitational influence. We might expect that some very large, massive clouds of gas and dust would condense into star clusters—and they do! The Pleiades, also known as the Seven Sisters, are a good example of what astronomers call an open star cluster. On a clear night, most people can count six stars in the Pleides with their unaided eyes. Those with sharp vision can count seven stars. But with binoculars or a small telescope, it becomes immediately evident that there are far more than seven stars in the Pleiades.

Open clusters are also called galactic clusters, because they are almost always found near the plane of the Milky Way. The open clusters are loosely assembled, and do not show a very striking concentration toward their centers.

The interesting thing about star clusters is not particularly unexpected: Most of their members form at about the same time. This helps astronomers to learn how stars evolve, because all the stars in a cluster are of about the same age. Yet, the more massive stars in a cluster appear to show the signs of old age even while the smaller stars are still young. A cross-sectional sampling of the stars in an open cluster can be plotted on a Hertzsprung-Russell diagram, and when this is done, the points on the chart are congregated quite close to each other. Very interesting patterns are often seen.

How do we know that the stars in an open cluster are actually associated, and are not just accidentally close to each other as we see them? The investigation is carried out by finding the radial velocities of the different stars in the cluster. We would expect that the stars in a cluster, affected by a common gravitational bond, would be moving together. This is what we indeed find; their radial speeds are all pretty much the same, and if the stars were not associated, we would find differences. Clusters tend to stay together because they move together.

The most spectacular star clusters are the globular clusters, which often have more than 100,000 stars. Globular clusters are so named because of their spherical or oval appearance; a good example of this is the cluster M13 in the constellation Hercules (Fig. 2-21). Over 100 globular clus-

ters have been observed in our galaxy, and there are almost certainly many more that we cannot see through obscuring dust clouds. The globular clusters, unlike the open clusters, seem to arrange themselves away from the galactic plane, in a large spherical halo around the Milky Way. The reason for this odd distribution is not fully known.

In the most dense central region of a typical globular cluster, the stars are about 20 times closer together than they are in the vicinity of our own sun. Imagine what it would be like if the sun were a member of a globular cluster such as M13, which is estimated to have about a half million individual stars! The night sky would surely be a spectacular sight.

The stars in a globular cluster are often bright giants and supergiants. Globular clusters contain an unusual preponderance of RR Lyrae type variable stars. If we dwelt in a globular cluster, we would be able to see hundreds of obviously variable stars in our sky. We would surely have learned about such stars much sooner. But perhaps our general knowledge of astronomy would be stunted by the overpowering luminosity of the cluster. We would probably not be able to see the rest of the Milky Way, although this would depend on our location with respect to the cluster and the galaxy. We might not ever discover that we live in a spiral galaxy, of much greater proportions than the cluster itself. The constant daylight would make telescopic observation nearly impossible until we acquired the ability to build telescopes in space, above our atmosphere. But then, this very process might be delayed by the lack of visibility from the ground! We do not, of course, know for certain what would have happened if our planet had been in such a cluster.

In 1917, the astronomer Harlow Shapley constructed a three-dimensional model, based on his observations, of the globular clusters with respect to our sun. He found that the clusters are arranged in a spherical shell about 100,000 light years across, and that the sun is fairly close to one surface of this sphere. Our system of stars seemed to have about the same diameter, 100,000 light years. Shapley sought to find whether the stars continue indefinitely into space, or whether they are con-

Fig. 2-21. The globular cluster M13 in Hercules has hundreds of thousands of constituent stars (courtesy of Palomar Observatory, California Institute of Technology).

fined to a certain region. Astronomers first began to recognize the galaxy as a spiral-shaped congregation of stars during the early part of the twentieth century. It seems odd, sometimes, to think that Albert Einstein was well along with his general theory of relativity before we knew what the spiral nebulae were, and before it was finally established that the Milky Way is one such nebula.

THE MILKY WAY

In the northern hemisphere, the Milky Way presents itself to us on summer evenings as a faint, hazy band of light, arching down from near the zenith to the northeast and southwest horizons. In winter, the Milky Way again appears high in the sky in the evening, this time passing from northwest to southeast. It is difficult to see the Milky Way from any vantage point near the lights of civilization; but well out into the country, it can be observed easily. The center of our galaxy, the brightest part of the Milky Way in the sky, is near the constellation Sagittarius. Figures 2-22 and 2-23 are photographs of our galaxy as seen in that direction.

Through a telescope or a pair of good binoculars, the hazy glow of the Milky Way is resolved in spectacular detail. Dark rifts often appear in time-exposure photographs, breaking the band of stars. Some of these dark regions appear rather like ragged storm clouds, and are called diffuse nebulae. As we look toward the center of the galaxy, as in Fig. 2-23, we can see much of the interstellar gas and dust, obscuring the light of the stars behind it. It was only recently that this dust was recognized for what it really is—an opaque rift in the galaxy—instead of simply a diluted region of stars. Most of the diffuse nebulae in the galaxy are concentrated near the plane of the disk, and since we are near the edge of the disk, we see the most nebulae as we gaze toward the center of our galaxy.

The resemblance of the dark nebulae to storm-tossed clouds goes further than simple visual similarity. Radiation from the stars in the galaxy, and the general movement of the stars around the center of the Milky Way, blows the nebulae around in great billows and vortices. The atoms of the diffuse nebulae are affected not only by radiation, but by magnetic fields and gravitational influences. Some of the vortices become tighter and tighter under the force of their own gravity, like hurricanes strengthening in the tropics of the earth. There is nothing to stop the clouds from collapsing upon themselves, and it is from these systems that stars are born.

Once astronomers gained access to large telescopes and sophisticated electronic equipment, a visual world was opened to them that revealed many things, including strange, glowing clouds in the interstellar void. Such clouds are called emission nebulae, since they appear bright against the dark background of space. If you look closely at the center star in the sword of the constellation Orion, for example, you might notice that it seems to have a blurred region around it. With a pair of large binoculars or a small telescope, the hazy patch becomes more well defined, and you can be certain that it is not just an optical illusion. When photographed through a large telescope, the hazy patch achieves a size of about 1 angular degree, or twice the apparent diameter of the sun or moon! This glowing, wispy mass, known as the Great Nebula in Orion, was one of the earliest recognized emission nebulae (Fig. 2-24). This cloud of interstellar gas and dust, ionized by the ultraviolet radiation of nearby stars, is about 1600 light years from our solar system, and has an actual diameter of about six times the distance from the sun to the nearest star.

Emission nebulae can take on almost any shape. One of the more interesting of these is the Ring Nebula in Lyra, shown in Fig. 2-25. There are many other such ring-shaped emission clouds visible in our galaxy through the telescope. Some are quite large, and others are so small that they can be differentiated from ordinary stars only by the peculiarities of their spectra. It would seem that these clouds should be more than just glowing doughnuts in space! In fact, it is believed that the ring-shaped nebulae are really spherical, and that we see them as rings simply because we see more material through the outer edges than through the center. (This happens for the same reason we see stars poorly near the horizon, when their light must

Fig. 2-22. The most dense part of our galaxy is in the direction of the constellation Sagittarius (courtesy of Mt. Wilson and Las Campanas Observatories, Carnegie Institute of Washington).

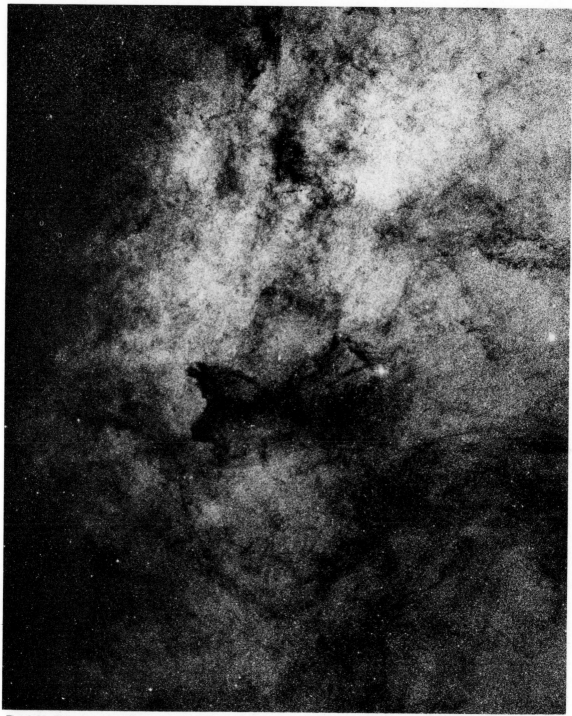

Fig. 2-23. This view of our galaxy in the region of Sagittarius was taken with the 48-inch telescope (courtesy of Palomar Observatory, California Institute of Technology).

Fig. 2-24. The Great Nebula in Orion is 26 lights years across (courtesy of Mount Wilson and Las Campanas Observatories, Carnegie Institute of Washington).

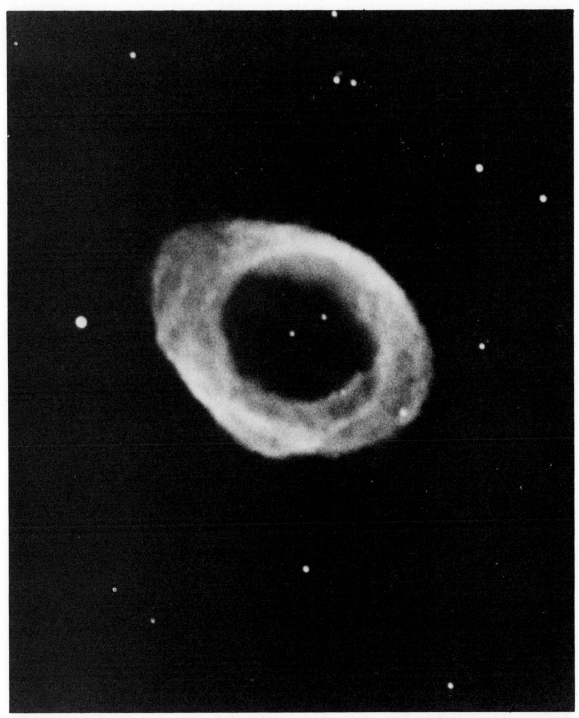

Fig. 2-25. The Ring Nebula in Lyra is one of many so-called planetary nebulae visible with large telescopes (courtesy of Palomar Observatory, California Institute of Technology).

pass through a large amount of atmosphere to reach us.)

Ring nebulae are sometimes called planetary nebula since, through a small telescope, they look disk-shaped. An amateur astronomer might even mistake a ring nebula for a planet! Several hundred such nebulae are known. There is invariably a star near the center of a ring nebula, and astronomers have found that they are all expanding. They can tell this from their spectral lines, which appear shifted toward the violet near the center and not toward the edges. This fact seems to imply that, at some time in the past, the cloud was thrown away from the star at the center. It could have been a supernova.

Dark nebulae, obscuring the stars behind, are seen not only in the Milky Way, but in other spiral galaxies. Since most of the dark nebulae are concentrated near the spiral plane of a galaxy, we see the clouds best when we view a galaxy from nearly edgewise. We are, of course, looking at our own galaxy this way, from very close range! Figure 2-26 shows an example of a view of a distant galaxy edge-on; this is what we might see if we could look at our own galaxy from a distance of several million light years. In this spiral, located in the constellation of Coma Berenices, dark nebulae appear distributed plentifully throughout the plane of the disk. There are many other such galaxies.

If there were no dark nebulae in our galaxy, or if the sun were situated far away from the galactic plane, we would see many more stars toward the center of the Milky Way. In fact, the central region of our galaxy, visible on spring and summer evenings in the constellation Sagittarius, might appear almost as bright as the sun itself. We would then have no darkness during much of the year! In some kinds of galaxies, there are essentially no dark nebulae. On planets in such galaxies, as on planets in or near globular clusters, life would be much different than we know it on the earth, with our constantly defined days and nights.

One of the greatest problems facing astronomers is the determination of the exact shape of our own galaxy. In particular, it is very different to get a good idea of the angle of the spiral arms with respect to geometric circles around the center. There

is no doubt that our galaxy is, in fact, a spiral galaxy; but the pitch of the spiral is not clear. It is very possible that the spiral is slanted more or less at the center than at the edges, and this compounds the problem. Our vantage point is too good!

The puzzle of the actual appearance of our galaxy, as it would be seen from an external point of view, has been attacked with radio telescopes as well as with optical instruments. Estimates of the pitch of the spiral, with respect to perfect circles, range from about 6 degrees to more than 25 degrees. In Fig. 2-27, geometric representations are shown for double concentric spirals with about 10, 20, and 30 degrees of pitch.

All of the spiral galaxies outside our own seem to have one characteristic in common: There are two spirals, emanating from opposite sides of the center. This is true no matter what the pitch of the spiral, and this can vary from almost zero to practically 90 degrees. We can be fairly certain that our galaxy has two spiral arms also.

ISLAND UNIVERSES

For a long time, the spiral nebulae were believed to be nothing more than emission nebulae within the confines of our Milky Way galaxy. Only in the early part of the twentieth century did astronomers deduce that these objects are actually other Milky Ways, some quite a bit larger than ours! Congregations of hundreds of billions of stars apiece, separated by vast tracts of dark emptiness, the galaxies have been called island universes.

Not all galaxies are spiral-shaped. Some of the largest and brightest galaxies are elliptical or spherical, resembling huge globular clusters. Elliptical galaxies are classified according to their eccentricity, which can range from spherical to more stretched-out than a football (Fig. 2-28). Elliptical galaxies contain very little diffuse matter, and it is believed that this is because all or most of the interstellar gas and dust have evolved into stars. Elliptical galaxies might therefore be expected to be old galaxies, with old stars, such as red giants and white dwarfs. While the white dwarfs in a distant galaxy are far too small and dim to be individually resolved, we do find many red giants. And

Fig. 2-26. See edgewise, a spiral galaxy shows gas and dust in the plane of its disk (courtesy of Palomar Observatory, California Institute of Technology).

giant they must be, indeed: Some of them are distinguishable as discrete points of light, in spite of their fantastic distance from us.

Some galaxies seem to have no defined shape whatsoever. These are the irregular galaxies. Our own Milky Way has two small irregular galaxies near it. These satellite galaxies are the Magellanic Clouds, named after the famous explorer who sailed around the world. The Magellanic Clouds are easily seen with the naked eye in the southern hemisphere. Some irregular galaxies do show signs of coordinated motions among their stars, such as a general rotation. Perhaps there is some degree of geometric definition in irregular galaxies, then; but visually it is not obvious.

The most stunning galaxies, from the standpoint of the visual observer, are certainly the spirals. Their variety is almost infinite. Some spirals appear broadside to us, some appear at a slant, and still others present themselves edgewise. The spiral arms can have many different shapes. The two main classifications of spiral galaxies are the normal spiral and the barred spiral. In the normal spiral galaxy, the arms extend from the bright nucleus, and are coiled in a more or less uniform fashion throughout the disk. Some spirals have prominent arms, while others have almost invisible arms (Fig. 2-29). In the barred spiral, the central region is somewhat rod-shaped, and remarkably straight. The spiral arms trail off from the ends of the rod, sometimes prominently, and other times almost invisibly (Fig. 2-30). The barred spirals are especially interesting, because the rod-shaped region appears to be revolving with constant angular speed at all points along its length. This seems contrary to the motion that should result in a rotating congregation of stars. Perhaps the nuclei of such galaxies are undergoing catastrophic explosions.

The mere appearance of a spiral galaxy gives the illusion that it is a rotating system. The rotation is fastest near the center of the normal spiral galaxy, and gets slower and slower toward the edge of the disk. This causes the spiral arms to trail behind. This is verified by spectral examination of different parts of galaxies viewed at a sharp slant. On one side of the galaxy, the light is shifted toward

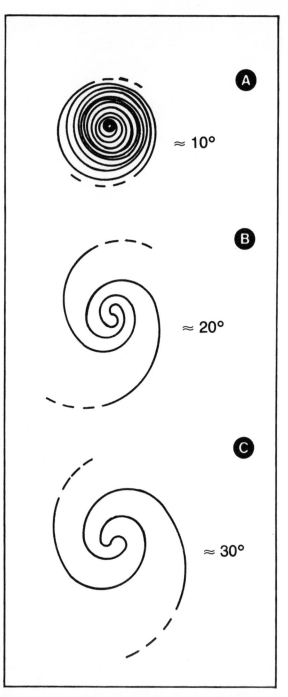

Fig. 2-27. Diagrams of spiral arms with various amounts of pitch. At A, 10 degrees with respect to the circular; at B, 20 degrees; at C, 30 degrees. The spiral arms of our galaxy are believed to have a pitch within the range shown here.

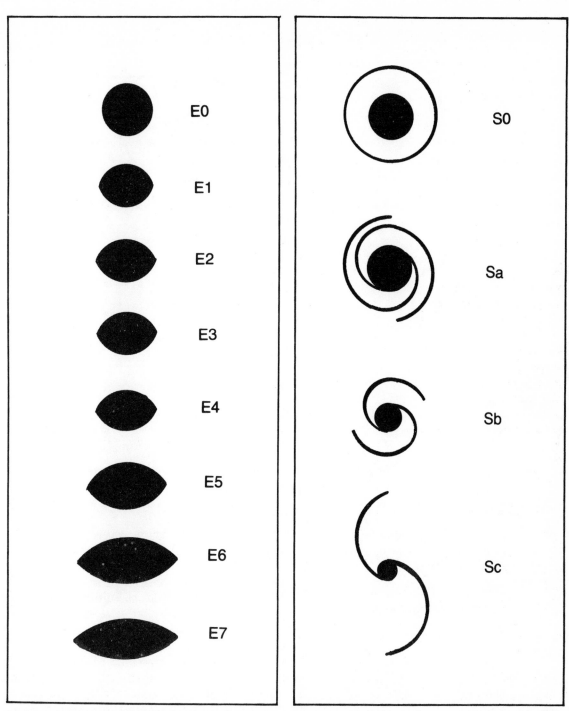

Fig. 2-28. Classification of elliptical galaxies, made by Edwin Hubble in the 1930s. The galaxy type E0 has a spherical shape. Type E7 has the greatest eccentricity.

Fig. 2-29. Hubble classification of spiral galaxies. Type S0 is shaped like an oblate sphere. The spiral arms wind less and less tightly for types Sa, Sb, and Sc.

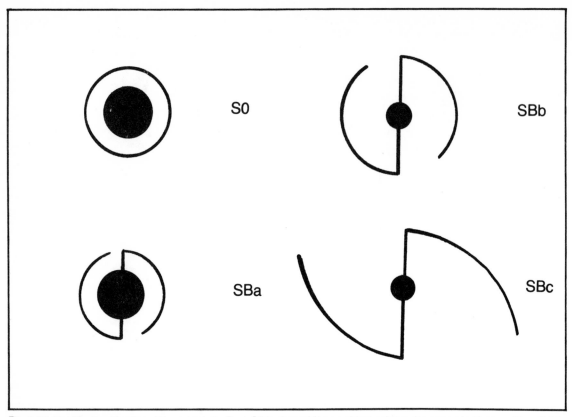

Fig. 2-30. Hubble classification of barred spirals. Type S0 is shaped like an oblate sphere; the arms wind less and less tightly for types SBa, SBb, and SBc.

the violet end of the spectrum, telling astronomers that it is approaching us. On the other side of the galaxy, a red shift indicates that the stars are moving away. Of course these determinations must be made independently of the overall motion of the galaxy with respect to us; in general, most galaxies are moving away, because of the expansion of the universe.

CLUSTERS OF GALAXIES

Our Milky Way galaxy has several close neighbors in space—close, that is, when we compare their distances to those of more remote galaxies. All of the galaxies in our "local group" are within 2 million light years of ours. The Great Nebula in Andromeda, called M31, is the most distant from us. The spiral galaxy M33, in the constellation Triangulum, is another member of the rather whimsically named "local group." There are several smaller irregular and elliptical galaxies as well. All of the galaxies in our cluster appear tilted at different angles in space.

When we look far into space, we find our galaxies in every direction. Along the spiral plane of our own Milky Way, of course, it is almost impossible to see exterior objects, because of the interstellar gas and dust that obscures the view. The farther out we look, the more galaxies we find. This is to be expected, since larger and larger spheres of observation will encompass more and more rapidly increasing volumes of space. But the distribution of the galaxies in the cosmos is, oddly, not altogether uniform. Clusters of galaxies are the rule, rather than the exception. Our "local group" is actually a small cluster; there are clusters with hundreds or even thousands of individual galaxies!

Just as stars can vary greatly in their character, and the galaxies can exist in many different and unique forms, so the clusters are found in differing shapes, sizes, and constitutions. The galaxies in some clusters are so close together that they collide often. After an encounter with another galaxy, a spiral may lose its arms because of the gravitational and electromagnetic disturbance. Or, it may become irregular and disorganized.

One of the most dense known clusters of galaxies is found in the constellation Coma Berenices. Near the center of this rich cluster, any galaxy can be expected to collide with another member several times during its life! What would it be like if our own Milky Way were currently in a collision with another galaxy? Such an event occurs over a period of millions of years. But we would certainly be able to tell if it were happening. There would be two sources of radio noise, not just one; and the noise level would probably be much greater than it is. If we had the opportunity to view the invading galaxy, it would probably present itself to us at such an angle that little interstellar material would intervene between its nucleus and our solar system. Then the nucleus of the invading galaxy would be plainly visible in our sky, and it might be quite bright—many times the brilliance of the full moon. The spiral arms, if the invading galaxy were a spiral, might be faintly visible.

How far into space do the clusters of galaxies extend? As far as we can currently see with optical telescopes, we can find galaxies. If we should ever be able to see to the edge of the universe with our telescopes, we might find strange and unknown types of galaxies. In fact, some unusually bright, small objects have been found at great distances, and can be seen with large telescopes. The radio telescope is the principal tool with which we can probe the limits of the cosmos, however, and we shall look at this electronic device in the next chapter.

THE EVOLUTION OF GALAXIES

We would naturally expect to find galaxies at different stages of their life cycles, since we can observe so many of them. But we know relatively little about how the galaxies actually were formed. It seems tempting to suppose that the galaxies formed from great whirling clouds of dust and gas, shortly after the big bang. But the puzzle is probably not that simple. Why do galaxies seem to occur in clusters, which themselves seem to show signs of rotation? Why are some galaxies so much larger than others? Why does it seem that the space between the galaxies is almost perfectly empty? Why do some spirals form at one tilt in a cluster, while others form at different tilts?

These questions are hard to answer, and there will no doubt be much more investigation into the riddle of the galaxies. In our attempts to understand the evolution process of the various kinds of galaxies, we can make use of one fortunate law of the physical universe: We see a galaxy, not as it is today, but as it was when its light left it. The farther into space that we look, the younger the image of the cosmos we see. If we are able to peer far enough, we might get a glimpse of the galaxies in their formative stages. We might see protogalaxies!

Our visual telescopes are limited by many complicating factors. It is hard to get detailed pictures of galaxies more than about 1 billion light years distant. A billion years must elapse for their light to reach us, and this is a long time, but not on the scale of stellar and galactic evolution. We need to see much farther back in time: 2 billion, 5 billion, even 10 billion years. It is here that the radio telescope has its greatest advantage over optical instruments. We can "see" all the way to the edge of the universe with the radio telescope, as Penzias and Wilson did in their discovery of the radiation from the big bang. We can "see" to that mystical boundary at which everything is flying away at the speed of light—so fast that we would never catch it, even with the fastest space ship. Beyond the edge of the universe, about 20 billion light years away according to the most recent estimates, we will never see anything. But 20 billion light years is far enough: The age of the cosmos is just 20 billion years!

The puzzle of galactic evolution may be solved, in part, by studying the strange objects called quasi-stellar radio sources, or quasars. First recognized by astronomers using radio telescopes, these visually unspectacular objects could hold a key to many of the mysteries of the universe. Let us see what has been found with the instruments of invisible astronomy.

Chapter 3

Quasars, Pulsars, and The Electromagnetic Spectrum

OBJECTS IN SPACE EMIT ENERGY IN MANY DIF-ferent forms. Visible light is, to us, immediately detectable, and for thousands of years it was the only vehicle by which astronomers could view the universe. The optical telescope was invented long before scientists knew that visible light waves are just a tiny part of a vast continuum of electromagnetic-energy wavelengths. It did not occur to our ancestors that the sun, planets, stars, and galaxies might emit radio signals, X rays, or gamma rays; such forms of energy were not known to exist.

Today, we realize that the range of electromagnetic energy is a wide realm, and we call it the electromagnetic spectrum. Astronomers can view the universe at an almost unlimited number of different wavelengths within this spectrum. The universe presents different "faces" to us at different electromagnetic wavelengths, and by studying these different pictures of the cosmos, astronomers are gaining a better understanding of the universe in which we, on our little satellite of a star, reside.

ELECTROMAGNETIC FIELDS

The nature of light was unknown to scientists for many centuries. Only speculation was possible. Isaac Newton was one of the first to provide a definitive theory of the nature of light: He believed it to be composed of tiny particles or corpuscles. Today, we recognize these particles as photons. But light is more complex than just a stream of particles. In the latter part of the nineteenth century it became apparent that light had wavelike properties. Under the right experimental conditions, light behaves like the ripples on a pond.

The wave action of light, and of other forms of radiant energy, is the result of a combination of the action of electric and magnetic forces. Charged particles, such as electrons and protons, produce electric fields. Magnetic poles produce magnetic fields. The fields extend into the space surrounding the charged particles or magnetic poles, and when the fields are strong enough, their effects can be noticed at a considerable distance. (The magnetic field of the earth affects a compass needle over the whole surface of the planet!) You have certainly noticed the attraction between opposite poles of magnets, and the repulsion between like poles. You may have noticed the same effects with electric fields. While these forces operate over just a short distance under laboratory conditions, this is only because such fields rapidly weaken, as the distance between poles increases, to less than the smallest amount we can readily detect. Actually, the magnetic fields or electric fields extend into space indefinitely.

Physicists have known for some time that a current in a wire will produce a magnetic field around the wire, perpendicular to the direction of the current. And it is well established that the mere existence of a voltage difference between two nearby objects will produce an electric field parallel to the gradient of the charge differential. But a changing current in a wire, or a changing charge differential between two nearby objects, will produce both magnetic and electric fields, in a combination that has a unique propensity for travelling through space. The electric and magnetic fields in such a situation are perpendicular to each other everywhere in space. The direction of travel is perpendicular to both the electric and magnetic lines of force, as shown in Fig. 3-1. Such a field is called an electromagnetic field.

In order for an electromagnetic field to exist, the electrons in a wire or other conductor must not only be set in motion, but they must be accelerated. That is, their speed must be made to change. The most common method of creating this sort of situation is the application of an alternating current to a length of wire. This is how a radio transmitter works. The frequency of the electromagnetic field is therefore the frequency of the alternating currents in the wire. This frequency may be very low, such as just a few cycles per second, or hertz (abbreviated Hz); the frequency may be thousands, millions, billions, or even trillions of hertz. Electromagnetic waves at different frequencies show much different properties. A frequency of 1,000 hertz is called 1 kilohertz (kHz); a frequency of 1,000 kHz is called 1 megahertz (MHz); a frequency of 1,000 MHz is called 1 gigahertz (GHz); and a

frequency of 1,000 GHz is called 1 terahertz (THz).

Electromagnetic waves travel through space at the speed of light, or about 300 million meters per second. The wavelength of an electromagnetic field therefore gets shorter as the frequency becomes higher. At 1 kHz, the wavelength is 300 kilometers, or about 188 miles. At 1 MHz, the wavelength is 300 meters, about three times the length of a football field. At 1 GHz, the wavelength is 300 millimeters, or about a foot. At 1 THz, an electromagnetic signal has a wavelength of a paltry 0.3 millimeters—so small that you would need a magnifying glass to see it. But the frequency of an electromagnetic wave can get much higher than 1 THz; some of the most energetic gamma rays have a wavelength of 10^{-5} angstrom units, or just one quadrillionth of a meter. The angstrom unit is equivalent to 10^{-10} meter, and is a standard unit to measure the wavelengths of electromagnetic disturbances at visible-light wavelengths and shorter. A microscope of great magnifying power would be needed to see an object with a length of 1 angstrom unit.

The discovery of electromagnetic fields led to the "wireless" radio, and ultimately to the sophisticated and complex variety of communications systems we know today. But radio waves are not the only form of electromagnetic radiation. Infrared, or heat, energy is of this nature. So is visible light, ultraviolet radiation, X-ray energy, and gamma-ray energy.

THE ELECTROMAGNETIC SPECTRUM

The wavelengths of very-low-frequency radio transmissions extend for miles. The shortest gamma rays have, as we have already mentioned, wavelengths that are only a tiny part of an angstrom unit. In between these extremes lie all of the different forms of radiant energy. To illustrate the range of electromagnetic effects, scientists often use a logarithmic scale based on frequency or wavelength. It is necessary to use a logarithmic scale since the range is so great, in terms of frequency or wavelength, that a linear scale would be hundreds of miles long! Figure 3-2 is such a logarithmic scale, based on wavelength. Each divi-

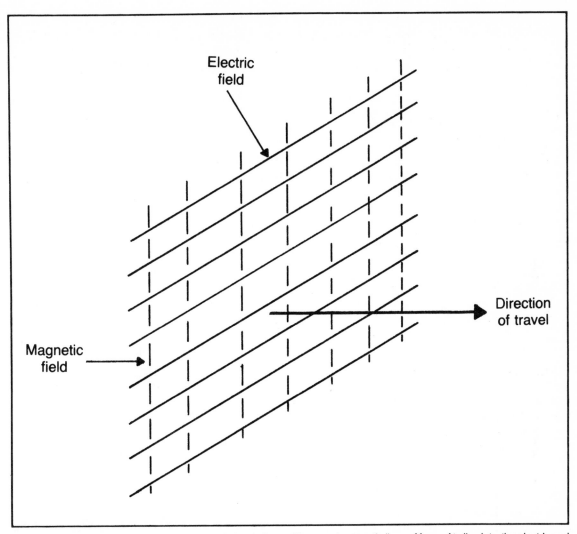

Fig. 3-1. An electromagnetic wave is made up of electric lines of force and magnetic lines of force. At all points, the electric and magnetic fields are perpendicular. The waves travel at right angles to both the electric and magnetic flux lines.

sion, in the direction of shorter wavelength, represents a tenfold decrease. The longest radio waves are at the bottom of the vertical scale, and the shortest gamma rays are at the top. Orders of magnitude are shown; the quantity 10^3, for example, indicates a 1 followed by three 0s, or 1000; and 10^{-3} indicates 0.001, or 1/1000. From this example, it is easy to see that visible light represents just a tiny part of the whole electromagnetic spectrum. It would certainly seem worthwhile to investigate the universe at other wavelengths!

To get some idea of what a tiny "window" to the cosmos is represented by the visible-light wavelengths, try looking through a red or blue colored piece of glass or cellophane. Such a color filter greatly restricts the view you get of the world; only a narrow range of visible wavelengths can pass effectively through the filter. Different colors cannot be ascertained through the filter. Through a red filter, blue appears the same as black, and red appears the same as gray or white. Other colors simply look like various shades of red, but there is

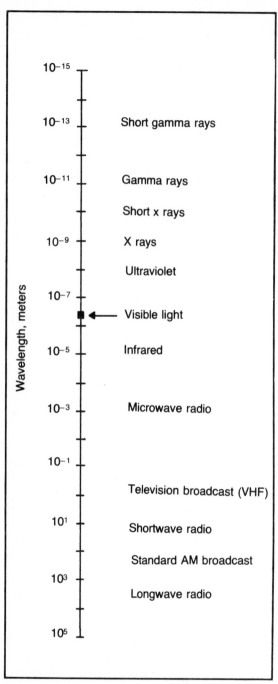

Wavelength, meters

10^{-15}	
10^{-13}	Short gamma rays
10^{-11}	Gamma rays
	Short x rays
10^{-9}	X rays
	Ultraviolet
10^{-7}	Visible light
10^{-5}	Infrared
10^{-3}	Microwave radio
10^{-1}	
	Television broadcast (VHF)
10^{1}	Shortwave radio
	Standard AM broadcast
10^{3}	Longwave radio
10^{5}	

Fig. 3-2. The electromagnetic spectrum, ranging from a wavelength of 100 kilometers to a quadrillionth of a meter. The visible spectrum is just a miniscule part of this range.

no color variety. Of course, if our eyes had built-in red color filters, our view of the world would be much different! But in an overall sense, all optical instruments suffer from just this handicap. We can see only a very small part of the whole electromagnetic spectrum; it is as if we were constantly looking through a color filter. The range of wavelengths we can detect, with our eyes, is about 7500 angstroms at the longest, and 3900 at the shortest. The longest wavelengths appear red to our eyes. The shortest appear violet. The intervening wavelengths show up as orange, yellow, green, blue, and indigo. To "see" below 7500 angstroms or above 3900 angstroms, we need special apparatus.

The observation of the universe at the visible wavelengths is not greatly hindered by the atmosphere of the earth. Some diffusion and refraction occurs, but the air is essentially transparent at the range of 7500 through 3900 angstroms. At some other wavelengths, however, energy from the cosmos cannot make it to the ground, and so we cannot detect it unless we go high up into the atmosphere, or even into space. At radio wavelengths greater than a few meters, the ionosphere, at levels ranging from about 40 to 250 miles, causes reflection of electromagnetic energy. This keeps signals from space away from us. (It is this effect, also, that keeps shortwave signals near the earth, and makes long-distance communication possible on the radio bands.) At ultraviolet wavelengths, the oxygen in the air is highly absorptive. In the X-ray and gamma-ray parts of the electromagnetic spectrum, our atmosphere completely obscures our view of the universe. But large portions of the invisible spectrum are observable with ground-based equipment, especially the wavelengths between the high-frequency radio band and the visible-light band. The most significant development in the exploration of space at the invisible wavelengths has been the radio telescope. Radio astronomy has become a refined science only since the development of wireless communications equipment, and, in particular, since the end of the second world war. High-altitude rockets, and the advent of the space age, have opened the newer fields of ultraviolet, X-ray, and gamma-ray astronomy.

THE RADIO TELESCOPE

Radio astronomy had its beginnings, as do so many scientific pursuits, in a sort of accident. Karl Jansky was conducting investigations at a wavelength of 15 meters, or about 21 MHz, in the high-frequency radio band, to determine the directional characteristics of atmospheric static. The antenna used was not very large; some of the high-gain Citizen's-Band radio antennas are almost as big. But Jansky found, in addition to the noise caused by local and distant thunderstorms, a weaker and steady noise of unknown origin. Man-made noise was ruled out when Jansky noticed that the source of the faint noise seemed to change with the time of day. It was found to have a rotational period of 23 hours and 56 minutes; exactly the same as the sidereal, or star-based, rotation time of the earth. Therefore, Jansky concluded that it was of extraterrestrial origin, and he found that it was coming from the direction of the constellation Sagittarius—from the center of the Milky Way galaxy. Other parts of the galaxy also produced radio noise, Jansky found, but not to the amplitude of the noise coming from the central region.

Unfortunately, although Jansky was interested in the phenomenon, and wanted to continue the research in the field with equipment designed specifically for receiving signals from space, his superiors were not as impressed. As a result, Karl Jansky did not pursue radio astronomy any further. He died at the relatively young age of 44, but his discovery of the noise coming from the Milky Way did not pass entirely unnoticed. A radio engineer named Grote Reber began to get interested in radio astronomy as a hobby, in conjunction with his activities as an amateur radio operator. Radio amateurs, also called "ham" operators, have had a long-standing reputation for making new discoveries in electronics and communications, and Grote Reber certainly is one of the more noteworthy.

Reber decided to build a large parabolic dish antenna in his own back yard. His neighbors were amazed as the assembly of the 31-foot reflector progressed; no doubt they wondered whether Reber had extraterrestrial connections! The antenna was not fully steerable, but could be moved only up and down, along the celestial meridian, from the southern horizon through the zenith to the northern horizon. As the earth rotated on its axis during the course of a day, different parts of the observable sky would pass across the focal axis of the antenna. Many radio telescopes use this kind of steering system. By tilting the antenna from horizontally south through the zenith to horizontally north, the entire sky can eventually be observed. Of course, the right time must be chosen to make an observation of a certain part of the heavens.

Reber's first tests were conducted at the fairly short wavelength of 9 centimeters, or a radio frequency of 3.3 GHz. He checked the most familiar objects in the sky, such as the sun, the moon, and the planets. No signals were detected. At a wavelength of 1.87 meters, or about 160 MHz, however, Reber did find noise coming from the Milky Way. Astronomers finally took notice of the work of Karl Jansky and Grote Reber in about 1940, and plans were made to construct large radio antennas to receive signals from the cosmos.

The most important part of any radio receiver is the antenna. This is especially true of the radio telescope. Radio signals from space are much fainter than standard broadcast or shortwave signals, where simple wire antennas are satisfactory. To determine the location in the sky from which a signal is arriving, it is necessary that the antenna be highly directional: It must be sensitive only to signals in certain orientations, and must be able to reject signals coming from unwanted directions. The television receiving antenna, which we often see on home and business rooftops, is a directional type of antenna, but the radio telescope requires a much more directive antenna than this. The gain, or sensitivity, of an antenna, as well as its directional resolving power, depend on the physical size of the antenna. For a given amount of gain and resolution, a dish antenna must be of a certain diameter, measured in wavelengths. Thus, high values of sensitivity and resolving power can be obtained more easily at the shorter wavelengths than at the longer wavelengths.

Of course, even the most sensitive and directional antenna is useless without a good receiver. The noise that comes from space sounds very much like the noise that is generated inside a radio receiver, and this compounds the problem of radio reception from the cosmos. (Tune your AM radio receiver to a frequency where there is no station. The faint hiss is internal noise; this is what radio astronomers generally hear from space.) The most advanced receiver designs must be used in a radio telescope, to obtain the greatest possible gain in the receiver.

The location of the radio-telescope antenna is important, just as is the site for an optical telescope. Manmade interference can completely ruin the operation of a radio telescope. Such interference comes from all kinds of electrical appliances, such as hair dryers, electric blankets, and thermostats. Automobile ignition systems are also a severe problem for those who attempt radio reception of faint signals. A country location is therefore vastly superior to an urban site for a radio telescope.

With all of these factors in mind, scientists set out to build the first genuinely sophisticated radio telescopes. One of the most famous of these first instruments was the 250-foot steerable dish located at Jodrell Bank in Cheshire, England. This project, completed in the 1950s, was proposed and overseen by the physicist A. C. B. Lovell. He went through great difficulties in arranging the construction of this radio telescope; at one point he feared that he might be put in prison for the financial difficulties the project encountered! Such is the dedication of scientists bent on unraveling the mysteries of the cosmos!

Not all radio telescopes use single, large dish antennas. There are other methods of obtaining precise resolution, and some are more practical than a huge parabolic reflector. The interferometer, pioneered by M. Ryle of Cambridge University and J. L. Pawsey of Australia, allows greatly increased resolving power using two separate dish antennas. When two antennas, spaced many wavelengths away from each other, are connected to the same receiver, an interference pattern occurs. There are many lobes, or directions in which the signals arriving at the two antennas add together; there are also many nodes, or directions from which the signals exactly cancel. The farther apart the two antennas, the more numerous the lobes and nodes; and they are very narrow, covering a much smaller part of the sky than the main lobe of any single practical antenna (Fig. 3-3).

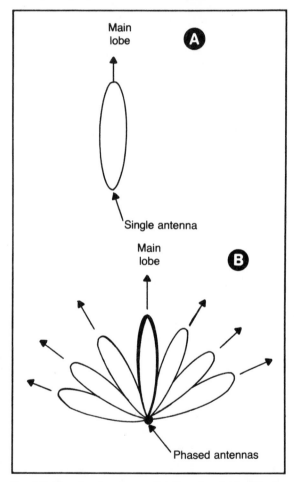

Fig. 3-3. The signal response direction of a single dish antenna might appear, if plotted mathematically, like the display at A. Two identical antennas, spaced several wavelengths apart, would then have a pattern like that at B; the main lobe is determined by the phase relationship of the signals arriving at the two antennas. This is how the interferometer works.

Another technique for getting fine resolution with a radio telescope is called aperture synthesis. It operates on a similar principle to that of interferometry. While interferometry and aperture synthesis cannot provide the sensitivity of a huge dish measuring miles in diameter, they do provide the equivalent resolution—and at a far lower cost and inconvenience! In some cases, the resolution can be on the order of just a few seconds of arc, or just a few thousandths of an angular degree.

Today, there are radio telescopes in many different countries throughout the world. The radio telescopes have certainly proved to be worth the trouble and the expense of their construction. The mysterious, fascinating quasars and pulsars were found using radio telescopes; only later did astronomers notice them with optical telescopes. Radio signals have been not only received, but transmitted using the large antennas, to aid in the mapping of the surfaces of Mercury, Venus, the moon, and Mars.

THE RADIO SKY

When a radio telescope with sufficient resolving power is used to map the sky, certain regions of greater and lesser radio emission are found. The center of our galaxy, located in the direction of the constellation Sagittarius, has already been mentioned as a powerful radio source. The sun is a fairly strong emitter of radio waves, as is the planet Jupiter.

A strong source of radio waves is found in the constellation Cygnus, and it has been named Cygnus A. Since the Australian radio astronomers J. Bolton and G. Stanley determined that Cygnus A has a very tiny angular diameter, and since they have found many other localized radio sources, a system of naming the radio sources has been adopted. A radio object is designated according to the constellation in which it is found, followed by a letter that indicates its relative strength within that constellation. The letter A is given to the strongest source in a constellation, the letter B to the second strongest, and so on. Thus Cygnus A is the strongest source of radio emissions in Cygnus; and not only in that constellation, it so happens—

Cygnus A is one of the loudest radio objects in the whole sky. It is so small in diameter that its output fluctuates because of effects of the earth's ionosphere as the signals pass through on their way to the surface. Cygnus A is called a radio galaxy.

Using radio telescopes, maps of the sky have been made, in the same way that optical astronomers make star and galactic maps. But radio maps do not look like the optical maps; instead, they appear more like topographical maps used in geologic surveys. Regions of constant radio emission are plotted along lines, which tend to be curved, as shown in the hypothetical illustration of Fig. 3-4. The finer the resolution of the radio telescope, the more individual, discrete radio objects can be defined on such a map. Invariably, in radio maps of the entire sky, the Milky Way shows up clearly, as a group of lines with their widest breadth in Sagittarius.

Other galaxies have been found that emit radio noise. Scientists at Cambridge University, in the early days of radio astronomy, identified four different external galaxies as radio sources. One of these was the Great Nebula in Andromeda, which is about 2 million light years from our own galaxy. Since the first discovery of radio emissions from other galaxies, it has been found that some galaxies are much louder than others. We will have more to say about this shortly.

RADAR ASTRONOMY

The huge parabolic antennas used in radio telescopes, as well as the other forms of high-gain antenna systems, will work for transmitting in the same way as they function for receiving. The large power gain developed in the interception of a faint signal from space will also greatly increase the effective transmitted power of a signal. By sending out short pulses of radio energy, generated by a transmitter in the laboratory, and then listening for possible echoes, it is fairly simple to get accurate determinations of the distances to other objects in the solar system. Amateur radio operators have heard the echoes of their own signals being reflected from the moon, and the antennas used for such communications have been quite small by

Fig. 3-4. A radio map of the heavens has contour lines, representing points from which equal-intensity signals are received. Each line represents a certain signal level. Here, a hypothetical discrete source is illustrated.

comparison to that of a radio telescope! The use of the radio telescope for distance determination, motion analysis, and surface mapping of extraterrestrial objects is called radar astronomy.

The distance to the moon has been refined to within a few inches, and it is possible to tell very precisely just how fast the moon is moving toward or away from us in its monthly orbit. The surface of the moon has been mapped by radar, and the quality of detail rivals that of optical photographs. Radar has proved to be a valuable tool in the study of the surface of Venus, because thick clouds make it possible to see the surface of the planet with optical telescopes. The main challenge to pursuing investigations of the planets has been the development of radar sets with enough power and sensitivity. Not only are the planets very distant, in terms of man-made radio transmissions, but they do not reflect the signals like a mirror would. Only a miniscule part of the signal reaching a planet is reflected back in the general direction of the earth; and of this reflected wave, only a very small part ever gets back to the antenna. The technology was eventually perfected to overcome this extreme path loss, however, and radar echoes from another planet were finally received. That planet was Venus. The radar telescope enabled scientists to discover that Venus has a retrograde motion on its axis—a fact that could not possibly have been ascertained with visual apparatus! Venus is unique in this respect. All of the other planets rotate counterclockwise as viewed from above their north poles; Venus rotates clockwise.

Radar mapping of the surface of Venus was attempted, since the signals easily penetrate the visually opaque clouds, and it was found that the ground is irregular. Although extreme resolution was not obtained, the discovery of mountains and plateaus on Venus points to a noteworthy fact: At least some of the surface is solid. This was not known with certainty until it was verified by radar. The Russian Venus probes have since, of course, taken pictures of the surface of the planet, and it indeed looks very solid and dry.

Observations of Mercury and Mars have also been made by radar. Mercury is very difficult to see with optical telescopes because it is so near the sun in the sky at all times. The rotation rate of Mercury was long thought to be 88 days, identical with its period of revolution around the sun. But the radar telescope showed that the rotation is actually completed in 59 days. Thus, Mercury does indeed have a day, although it is a long one.

Echoes from the planet Mercury have been used for a very different purpose, as well as surface mapping and the evaluation of its rotational period. This has been the verification of one of the predictions of Albert Einstein's general theory of relativity. According to the prediction, radio waves passing close to a massive object, such as the sun, should be slowed down somewhat. All radiant energy, according to this famous theory, slows down near a large mass because of the effects of gravitation. The radar telescope provided scientists with a means of checking this, by bouncing signals off the planet Mercury as it passed on the far side of the sun (Fig. 3-5). According to the prediction of relativity theory, Mercury should appear to deviate about 40 miles outside of its orbit. The apparent increase in the orbital radius of Mercury, as it passes on the far side of the sun, is exaggerated by the drawing, but the deviation was found. The echoes were slightly delayed as the signals passed near the sun on their way to and from Mercury. The magnitude of this discrepancy was found to be in close agreement with the predicted value according to the general theory of relativity.

Another application of radar astronomy is the study of the paths and orbits of meteors. Meteors apparently come, for the most part, from inside the solar system, and are not interstellar wanderers. The velocities of the meteors as they enter the atmosphere can be accurately measured using radar, and from the velocities, their orbits around the sun can be determined. The meteors arrive during the day as well as at night, and the radar telescope can "see" them just as well in daylight as in darkness.

The sun itself has been observed by radar. The outward motion of subatomic particles has been detected, and this is called the solar wind. The surface of the sun appears at a different level in the

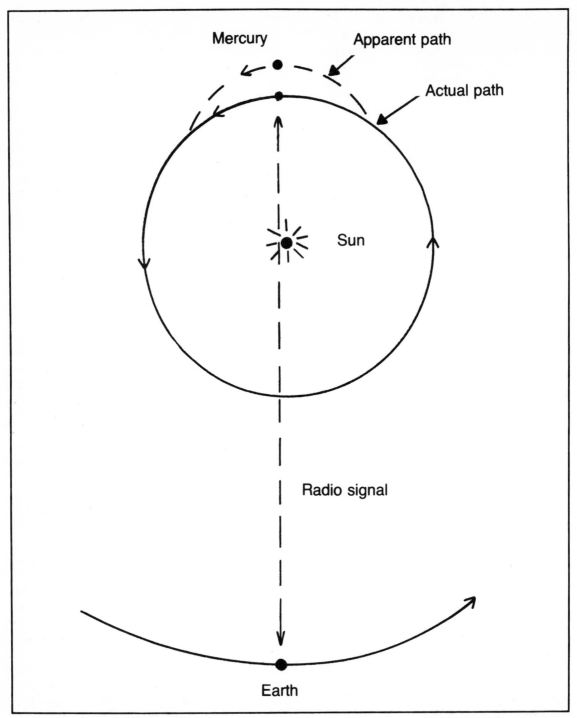

Fig. 3-5. When radar signals are passed very near the sun, they are slowed down, as predicted by the general theory of relativity. This makes the planet Mercury appear to deviate outside its orbit.

radio part of the spectrum, as compared with the visible part of the spectrum, and what is more, the apparent surface depends on the wavelength. This lets the radar astronomer examine the motion of the gases at different levels in the atmosphere of our parent star. Great turbulence exists there, because Doppler shifts are constantly observed.

The maximum effective range of the radar telescope is limited by two things. First, there is the problem with path losses, which mandates more and more sensitive receivers and more powerful transmitters as the distance gets greater. There is a limit to how sensitive a receiver can be; there is a limit to how much power we can generate in a transmitter; and there is a limit to how much gain we can get with an antenna. Technology still leaves some room for improvement in this respect. So we have some control, yet, over these problems. But the second limiting constraint is insurmountable: The speed of radio waves is finite. They always travel at just 186,282 miles per second. An echo from Pluto, the outermost planet (that we know of) in the solar system, would return after 10 hours. An echo from the nearest star system, the Alpha Centauri system, would not return for almost 9 years! That is a long time to wait. Most of the star systems in our galaxy are so far away that the echoes from a radar set would not come back until many human lifetimes had passed. So it is very doubtful that radar astronomy will ever be useful outside of the solar system.

This distance limitation is not too much of a problem for those astronomers who continue to be puzzled by the things in our own solar system—and there are certainly enough mysteries close to home, as well as far away.

RADIO RECEPTION FROM THE SOLAR SYSTEM

With the birth and maturation of radio astronomy, scientists turned their attention to several objects in our own solar system. One of these is the sun. The radio sun, as seen at radio wavelengths, is somewhat larger than the visible disk, and it appears oblate or flattened along the plane of the equator. The apparent diameter of the sun is less through the poles, and is largest through the equator.

Visible solar flares are also observed with the radio telescope. Such flares have long been associated with disruption of the ionosphere of our planet, a phenomenon that wreaks havoc with shortwave radio communications. There are several different kinds of solar flares at radio wavelengths. Radio outbursts from the sun usually portend a geomagnetic disturbance an hour or so afterwards, as the high-energy particles arrive and are focused toward the magnetic poles of the earth. Then, at night, we see the aurora. We also observe an abrupt change in radio propagation in earth-to-earth circuits at some frequencies.

Radio observations of the moon and planets have enabled astronomers to more accurately ascertain the surface temperatures, especially of planets with thick atmospheres such as the "gas giants" Jupiter, Saturn, Uranus, and Neptune. The inner planets are generally warmer than the outer ones, as we would expect. But Jupiter is the exception! This planet produces unusually strong radio emissions, and has a fairly high temperature deep within its shroud of gas. At a wavelength of about 15 meters, the electromagnetic radiation from Jupiter is almost as strong as that from the sun. The largest planet is also a very strong radio source at shorter wavelengths. Some of this radiation can be attributed to the fact that Jupiter, having almost become a star, generates considerable heat—far more than any of the other planets. But the internal heat of Jupiter cannot account for all of the radio emissions coming from the giant planet.

Several theories have been formulated in an attempt to explain the unusual levels of electromagnetic radiation coming from Jupiter. One theory holds that numerous heavy thunderstorms rage through the thick atmosphere, and that the radio noise is caused by the lightning discharges. But the noise is too intense for this idea to fully explain it. A more plausible theory is that electrons, trapped by the intense magnetic field of Jupiter, cause a form of electromagnetic emission called synchrotron radiation as they are accelerated along with the high rotational speed of the planet.

THE HYDROGEN LINE

Clouds of hydrogen gas are abundant in our galaxy, especially in the spiral arms. These clouds emit electromagnetic energy at numerous wavelengths in the infrared, visible, and ultraviolet ranges. The hydrogen clouds also have a characteristic emission at radio wavelengths, and this occurs at about 21 centimeters. The existence of this radiation was theoretically predicted by a Dutch astronomer, Henrik van de Hulst, and a Russian, Iosif S. Shklovskii, before it was actually observed with the radio telescope. It was surmised that this 21-centimeter emission would exist over a small band of frequencies, rather than at exactly one single frequency, because of Dopper effects resulting from the movement of the hydrogen clouds in the galaxy.

The 21-centimeter radiation was observed for the first time with radio telescopes in the United States, the Netherlands, and Australia. The gas clouds were located, and their motions were evaluated by observing the Doppler shifts in the emission wavelength. No doubt remained that our galaxy is indeed a spiral galaxy.

The 21-centimeter line is so unique, and so prominent, in the radio spectrum, that some scientists believe this frequency would be a logical choice for attempting to communicate with extraterrestrial civilizations. What better place in the spectrum, it has been argued, could exist for this purpose but a "cosmic marker" wavelength? Still, there are two sides to this kind of argument. Perhaps the 21-centimeter wavelength should be the one to *avoid*, since the background noise level is high in its vicinity, and this might tend to mask weak signals.

RADIO GALAXIES

All galaxies emit energy at the radio wavelengths, as well as in the visible range. Usually, the intensity of the radio emission from a galaxy can be predicted according to the classification of the galaxy and its observed visual brilliance. But some galaxies emit far more energy at radio frequencies than we would expect; these are the radio galaxies. The intense radio source Cygnus A in one such galaxy. When radio telescopes are used to map the details of Cygnus A, a double structure is found. The radio emission comes from two different regions, located on either side of the visible galaxy.

Other such double galaxies have been observed with radio telescopes. Several different hypotheses have been put forth in an attempt to find out what is taking place in such galaxies. One theory is that the radio galaxies are actually pairs of colliding galaxies. The magnetic and electric fields of the two galaxies might interact to produce unusual levels of radio-frequency energy. But there are certain problems with this theory. Isoif. S. Shklovskii of the Soviet Union has theorized that perhaps the radio galaxies contain an excessive number of supernovae, or exploding stars, as compared with normal galaxies. F. Hoyle and W. A. Fowler have suggested that the tremendous energy of the radio galaxies might come from explosions of the galactic nuclei, following or associated with a catastrophic gravitational collapse. The mechanism for such powerful radio emissions from these galaxies remains unclear. But, it is believed that the nuclei of the radio galaxies are undergoing radical changes. As more becomes known about radio galaxies, astronomers hope to further unravel the puzzles of galatic formation and evolution.

In 1960, the position of a strong radio source was defined to within a tiny fraction of an angular degree, and its size was found to be less than 1 second of arc in an angular sense. Comparing the position of this radio source with various visible objects in its vicinity, the strange "radio star" was found to be a rather faint blue star in the photographs. But there was something especially odd about this star: The astronomers J. L. Greenstein and A. Sandage could not identify the absorption lines in its spectrum. It did not, however, take them long to find the problem! The red shift in this source is so great that the spectral lines are greatly altered; the object is receding from us at a sizable fraction of the speed of light, and so the spectrum is drastically red-shifted.

Soon after the discovery of this "radio star," several other similar objects were found, and they also had very large red shifts in their spectral absorption lines. The objects, because of the visual resemblance to stars and because of their strong radio emissions, were called quasi-stellar radio sources. This name has since been shortened to the more palatable term "quasar." These objects are some of the most intriguing in the universe.

LOOKING AT QUASARS

After the first few quasars were found, many others were discovered and observed. In many cases, it was discovered that the quasars had been previously photographed many times; but in the photographs they had appeared as stars, and were dismissed as nothing but stars. In one case, when several photographs were examined having been taken over a period of decades, it was found that a quasar had fluctuated in brightness. Large changes had occurred within periods of just a few months.

Whatever the reason for the changes in the brilliance of the quasar 3C48, the first-discovered of the quasi-stellar radio sources, the fact that it could take place within a short period gives us a clue as to its diameter. Nothing can propagate at speeds greater than the speed of light; this is true of causes and effects, as well as of energy! Apparently, then, the quasar 3C48 is but a few light months across. This makes it a strange object indeed, because if its red shift is a correct indicator of its distance, its energy output is many times that of a normal galaxy! Quasars are evidently very concentrated, as well as immensely powerful, sources of energy.

Using radio telescopes with extreme resolving power—just a few percent of 1 second of arc in some cases—some of the quasars still appear as point sources of energy. This is also true of the nuclei of certain radio galaxies. Optically, all of the quasars look like point sources of light, and therefore they resemble stars, no matter how powerful the telescope. It is little wonder that the peculiar nature of the quasars went unnoticed until the radio telescopes were trained on them!

Some quasars can be resolved into compo-nents by means of the radio telescope, but this requires the technique of interferometry, and the use of a base line of hundreds or even thousands of miles. Separate antennas are linked by computer in order to accomplish this. The resulting angular resolution goes down to less than a thousandth of an arc second, or about ¼ of a millionth of an angular degree. With such sophisticated radio apparatus, the components or fringes have been seen within various radio galaxies and quasars.

There is another, quite different way to estimate the angular diameter of a radio source, and that is by observing rapid changes or scintillations in its brilliance as the radio waves pass through turbulent ionized clouds of interplanetary particles. You have certainly noticed the twinkling of the stars, even on clear nights; yet the planets appear to shine almost without blinking at all. The reason for this is that the planets have a much greater angular diameter than any star. Small telescopes show the planets as disks, but the nearest stars look like points of light, even in the most powerful telescope. Turbulence in the air, like the heat scintillations that rise from a hot pavement on a summer afternoon, cause a point source of light to twinkle as we watch it. The charged atomic particles that comprise the solar wind, as they stream outward from the sun, have an identical effect on radio signals coming from deep space. A source of radio waves with a small angular diameter seems to scintillate.

By observing the quasars with a single radio telescope, and carefully recording the intensity of the emissions reaching the antenna, it is possible to get an accurate idea of the angular size of a radio object. Quasars always appear as very small sources of radio energy, at most a few light years in diameter. Other observed properties of the quasars, such as curvature in the spectral lines, have led astronomers to believe that they are small, even though they emit such fantastic amounts of energy.

The sizes of the quasars, as well as the estimates of their energy output, have been determined according to the Hubble relation between red shifts and distances. All of the quasars show large red shifts in the absorption lines of their spectra, and this has led astronomers to surmise that they are billions of light years away from our own galaxy. But perhaps the red shifts are being misinterpreted; could it be that the quasars are really very tiny, local objects, and that they are being thrust outward from our galaxy at tremendous speeds? This seems unlikely, for at least two reasons. First, since our solar system is located near the edge of our galaxy, we would expect that objects ejected from the galaxy would, at least sometimes, approach us and not move away. This approaching speed would give such objects a pronounced violet shift, because of the Doppler effect. But not a single quasar has ever been found with a violet shift; they always have red shifts. Second, we might logically expect that if quasars are ejected from galaxies, we should see them coming from other galaxies near us, such as the Great Nebula in Andromeda. Some of the objects ejected from the nearby galaxies would have a red shift, as they were moving away from us; but some would also have a violet shift, and again, not a single quasar has even been seen with a violet shift. Some of the objects would be moving just about sideways to us, and they would have no shift; but no quasar has been seen with a zero spectra shift, either! Quasars, without exception, have very large red shifts.

A completely different attempt has been made to prove that quasars are local objects, and not distant, tremendous sources of energy. Albert Einstein showed, in the formulation of his general theory of relativity, that a very intense gravitational field can produce a red shift in the spectrum of the light coming from the source of the gravitation. This effect has been observed even in the comparatively weak gravitational field of our own planet, so we know that Einstein's theory is correct. If the quasars are very small, rather local, dense stars, then perhaps the red shifts in their spectral lines can be justified in terms of the relativistic effect of gravitation. An extremely dense object, with a powerful gravitational field, could produce a large red shift.

The first quasar that was discovered, 3C48, would have to weigh the same as the sun, be just 6 miles in diameter, and be in the troposphere of our planet, in order to account for the radiation intensity it possesses! This calculation was made by cosmologists, and when they found out the results, they were pretty much convinced that the quasar could not be very small nor very local! But even if 3C48 had thousands of times the mass of the sun, it would still have to be within the solar system, and this obviously cannot be the case, since such a massive object would be the center of revolution for all the planets. The derivations in these terms for other quasars give similar results. A tempting theory is dismissed.

In the case of the quasars, scientists have employed the "devil's-advocate" approach in an attempt to prove that they are distant and brilliant. The "devil's-advocate" scheme makes attempts to disprove a theory, and if these attempts fail, then the theory is lent support. The quasars, even after attack by the "devil's-advocate" scientific method of inquiry, appear to be very far-off and very energic cosmic phenomena. There seems to be no room for doubt.

WHAT ARE QUASARS?

The anatomy of the quasars is a great puzzle to astronomers, and the unraveling of this riddle is currently under way. It seems that the quasars are very distant, and also very powerful, sources of energy. Suppose we *postulate* that this is in fact true. If we are willing to postulate this, then there are certain things that can be deduced about the quasars, based on this axiom.

First, although hundreds of quasars have been identified, now that astronomers have learned how to recognize them, the quasars are much farther away than all of the nearby galaxies. (In this sense, "nearby" means within several hundred million light years!) Quasars are not only distant, but they are *extremely* distant. Without exception, they display large red shifts in their spectral lines. In the cosmos, distance is time; when we look at a galaxy that is 1 billion light years away, we are looking 1 billion years into the past. The quasars are, in general, all billions of light years from us, and therefore they are also billions of years back in time. As we look at a quasar, we are gazing into a past so remote that the earth itself was surely much different than we know it today. Could it be that the quasars are an almost universal phenomenon of the universe in its younger age? Astronomers find this to be a tempting idea. The present estimate of the age of our universe is about 20 billion years. Some of the quasars, at distances approaching 10 billion light years, are thus images of the universe at half its present age! Many stars have lifespans of much less than 10 billion years. The galaxies themselves are made up of stars. Perhaps quasars are very young galaxies.

Observations of quasars and radio galaxies often reveal striking similarities, and some astronomers believe that quasars and radio galaxies are in fact the same. The nuclei of the radio galaxies appear to have very small diameters, and yet they put out vast amounts of energy. This makes them very much like the quasars. It is possible that we see only the brilliant nuclei of the most distant radio galaxies (Fig. 3-6). The rest of the galaxy might be drowned out, visually, by the nucleus. Both the quasars and the radio galaxies, if they are young galaxies, are important clues in the pursuit of knowledge about the evolution of galaxies in general.

When large amounts of matter are concentrated into a small volume of space, the gravitational influence of the matter can have profound results. We will have more to say about this in the next chapter. For now, let us just take note that a dense congregation of stars, such as we might find in the center of a galaxy, might gravitationally seal itself off from the rest of the universe. The stars near the periphery of the congregation might orbit the central region at great speeds before being pulled forever into the mass. This high velocity, and the accompanying magnetic fields, would produce large amounts of electromagnetic energy at visible and radio wavelengths. A quasar may be an active *black hole*, a gravitational one-way membrane—literally a cosmic hurricane (Fig. 3-7)!

Yet another theory concerning the origin and anatomy of the quasars suggests that they are points in space through which new matter is entering, perhaps from another universe in another dimension of time and space. Such objects, in the imagination of the cosmologist, are called *white holes*. As matter bursts into our space-time continuum, having been pulled from another universe by overwhelming gravitational forces, the flash of radiant energy would surely be brilliant (Fig. 3-8). This idea, while somewhat far-fetched simply because of a lack of direct evidence, is one of the most fascinating in all of cosmology. Is it possible that there are other universes, with "wormhole" time-space singularities connecting them with our cosmos?

What are quasars? The question has not yet been completely resolved. Some day, when gigantic optical telescopes are built on the moon or aboard space stations, where no hindering air molecules limit the detail we can see, we might be able to watch a quasar in action. This should help astronomers describe just what they are, and how they operate and evolve. Observation at wavelengths both shorter and longer than the visible spectrum are also being conducted, and are proving invaluable.

LITTLE GREEN MEN

The phenomenon of scintillation, or rapid fluctuation, of radio sources has been important, as we have seen, in the estimation of the angular sizes of radio objects such as quasars. At the radio-telescope observatory of the Cambridge University in England, an antenna was specially assembled during the mid-1960s for the purpose of conducting an investigation of the scintillation of radio sources. The antenna was made up of more than 2,000 individual dipoles in a massive phased array covering more than four acres of ground. The frequency chosen for the experiments was about 81 MHz, which is just below the standard FM broadcast band and within the limits of U.S. television channel 3. This corresponds to a wavelength of 3.7 meters.

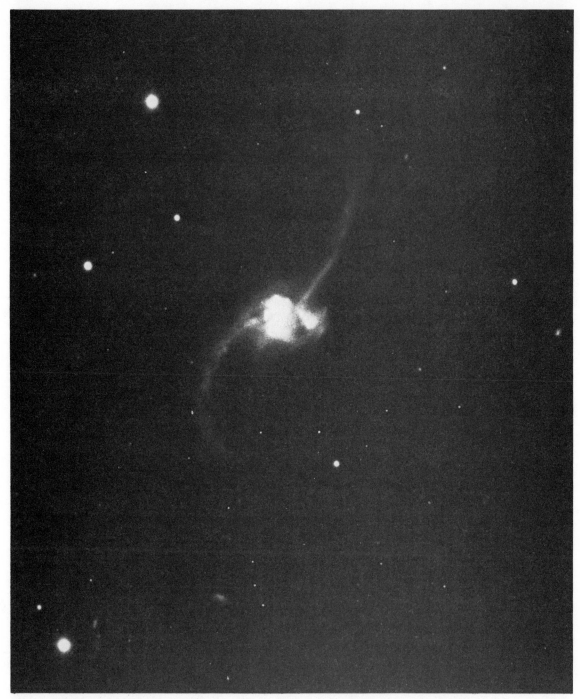

Fig. 3-6. Radio galaxies with extremely bright nuclei may appear, from a great distance, as small, bright objects, such as this galaxy, NGC2623, in Cancer (courtesy of Palomar Observatory, California Institute of Technology).

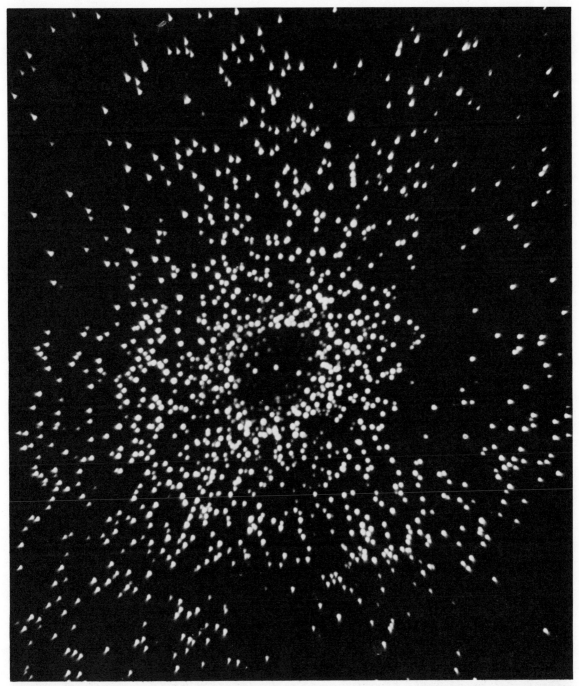

Fig. 3-7. A quasar may be a huge black hole. In this artist's conception of a black hole, thousands of stars are being pulled inward by the gravitational effects of the object, and this results in the emission of great amounts of energy at radio wavelengths.

Fig. 3-8. A quasar might be a point at which new matter is entering the universe—a white hole.

Because of the rather long wavelength chosen, extreme resolution and antenna steerability were not possible. The scintillations were, however, observed, and the research went well. But there were some unexplained radio signals, which could not fit into the overall picture. Jocelyn Bell and Anthony Hewish, a graduate student and her professor, noted these strange "scintillations" in their recordings, and designed special high-speed devices to examine them more closely. The new recorders, capable of responding to rapid changes in signal strength, were finished in 1967, and in the summer of that year, the results were obtained. The signals were so astonishing to the scientists that they did not immediately release their findings: They were hardly believable!

A radio source, having a defined position in the sky, was emitting sharp pulses at intervals of about 1 second. The regularity of the pulses was so uniform that the signal seemed as though it might be generated by intelligent beings. The duration of each pulse was just a few milliseconds. Could it be

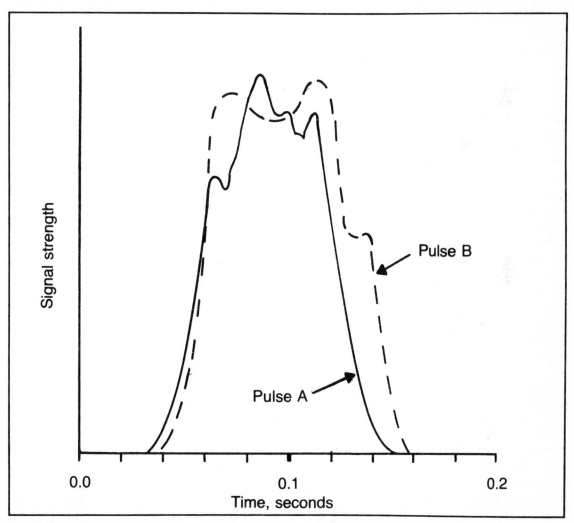

Fig. 3-9. The emission from a pulsar has an irregular shape, and it varies from one pulse to the next. Here, a pulse is shown by the solid line, and the following pulse, a certain time later, is illustrated by the dotted line. This is a hypothetical, but representative, situation.

that some little green men were broadcasting time signals, for some reason, into outer space? The sound of the signal was reminiscent of the terrestrial time-standard stations. The strange pulsing radio source was called a pulsar, and was given the designation CP1919, meaning "Cambridge Pulsar at right ascension 19 hours and 19 minutes."

Before the results of their discovery could be released, Bell and Hewish had to be certain that the signals actually were coming from space, and were not some odd form of earthly radio interference. The position of the pulsar did, however, remain at 19 hours and 19 minutes right ascension at all times, and was fixed with respect to the stars; the source came across the meridian once every 23 hours and 56 minutes, or exactly once each sidereal (star) day. Doppler shifts have since been noted in the period of the pulsar as the earth, in its circumnavigation of the sun, first moves toward CP1919 and then away from it.

For a time, the possibility was seriously entertained that the weird radio object CP1919 might be an artificially generated signal, produced by intelligent beings. But if these were so, we would expect to find a Doppler shift associated with the revolution of the alien planet around its sun; and no such shift was found. Thus the little-green-men hypothesis had to be abandoned. The pulses were seen to differ, in terms of their internal detail, from cycle to cycle; the "ticks" have an irregular shape that is not always the same (Fig. 3-9).

The measurement of the angular diameter of the pulsar CP1919 revealed that it is a very small object. This did not come as much of a surprise to astronomers. The brevity of the pulse period and, especially, the shortness of the pulses themselves, suggested a generating source that could be no more than a few hundred miles across. But what known astrophysical phenomenon could account for such strong signals from such a small source?

Estimates of the distance to CP1919 showed that it is about 400 light years away. This measurement was made using a truly ingenious formula. At progressively longer wavelengths, the pulses arrive at greater and greater intervals. This indicates

that the speed of the propagation of radio signals through space depends on the wavelength. This effect is not peculiar to pulsars; it occurs with the radiation from all objects. The difference in the speed of radio-wave propagation as a function of wavelength does not amount to much. Sophisticated instruments are needed to detect any difference. But it is a definite occurrence, and it takes place because of dispersion caused by free electrons in the interstellar medium. The greater the distance to a radio source, the longer the signals have been en route to our antennas, and the more dispersion will be observed. Knowing, or at least making an educated guess at, the concentration of free electrons in space, the distance to a radio source can be determined if the intensity fluctuates. A pulsar is very convenient in this respect! Its fluctuations are extremely great and precisely timed!

After the discovery of the pulsar CP1919 was announced by the Cambridge team, radio astronomers throughout the world began to look for more pulsars. Radio telescopes of the highest sensitivity and resolving power were put to the task. It was not difficult to find pulsars, once the radio astronomers knew what to look for. Early in 1968, the announcement was made that three more had been located, and since then, many pulsars have been catalogued. Not all pulsars have the same period or pulse duration. Some pulsars cycle once in only a few hundredths of a second, and others have periods of several seconds. The pulse lengths vary as well.

WHAT ARE PULSARS?

Theories concerning the anatomy of pulsars began to condense after Tommy Gold, one of the original proponents of the steady-state theory of the universe, hypothesized that the strange objects are fast-spinning neutron stars that generate intense electromagnetic fields. When a star has used up all of the fuel available for the process of nuclear fusion, gravitation takes over, and the star is compressed into a fantastically dense and bizarre form of matter. If the star is massive enough to begin with, the compressing force can get so great that the electrons are driven into the protons of the nuclei, ending up as neutrons. Such a situation results in all of the mass of the star being contained in a sphere having a diameter of only a few miles.

The key to the shortness of the pulsar cycle, which baffled the astronomers, seemed to be the angular-momentum effects on the rotational speed of a collapsing star. You have probably seen how a figure skater, when performing a spin, begins slowly, with outstretched arms; the skater speeds up the spin by pulling the arms inward toward the body. This reduces the effective diameter of the mass of the body; to conserve momentum, the body speeds up. (When properly executed, this is very dramatic, and it makes you wonder why the skater doesn't faint from sheer dizziness!) A collapsing star behaves in exactly the same way, but the increase in speed is much greater. The radio signals we hear are apparently the result of the combined effects of size reduction and relativistic behavior of the magnetic fields of the collapsing star. According to Gold, the magnetic field near a collapsing neutron star could grow to the unearthly intensity of trillions of gauss. The contracting magnetic field would squeeze the magnetic lines of flux into a smaller and smaller volume, and the tremendous gravitation and speed of rotation would cause a relativistic "piling-up" of the magnetic field upon itself. The result would be the generation of great bursts of electromagnetic energy, in the form of radio waves, infrared radiation, visible light, and even ultraviolet and X rays.

If Tommy Gold's theory is anywhere near correct, then it would be logical to conclude that the rotational periods of the pulsars should get gradually longer and longer. All spinning objects rotate more slowly as time passes, because they lose momentum. Careful observations of pulsar periods, taken over a long span of time, have confirmed that this does happen. The cycle of the pulsar invariably gets longer. Gold's theory is thus lent support, and all current theories of the pulsars are variations of Gold's hypothesis.

Sometimes the energy from a pulsar arrives in such intense bursts that we can see them with optical telescopes! A pulsar in the Crab Nebula, which is believed to be the remnant of a supernova explosion, has a period of about 0.03 second, and it emits visible flashes. Such a rapid rate of "flashing" went unnoticed for many years, since most visual astronomy is carried out by means of time-exposure photographs because the sources are so faint. Such time-exposure photographs would obscure the pulsing of a light source, and thus the star in the Crab Nebula was not seen as a pulsating light source until such effects were sought and found.

Pulsars remain very strange and mysterious. Although Tommy Gold's hypothesis, or variations of it, have been generally accepted as explaining the intense emissions in a qualitative way, no definitive model for a pulsar has ever been constructed. It is impossible to make a pulsar in the laboratory, and computer analysis is not conclusive because of the lack of adequate data. To further confuse the astronomer and astrophysicist, some pulsars have been seen to stop their emissions for up to several seconds or minutes, and then start again! Until more data is gathered from the pulsars, scientists cannot be absolutely certain what they are. An entirely new, and radically different, explanation may someday be found for the intense, extremely regular, electromagnetic emissions.

We can say one more thing about neutron stars that apparently holds the key to the phenomenon of the pulsar: Such highly compressed matter can behave in incredible ways. As a neutron star collapses because of its own gravitation, it gets so dense that the escape velocity—the speed needed to accelerate away from the star—approaches the speed of light. In some situations, the escape velocity can, according to the equations derived by astrophysicists, actually exceed the speed of light, and if that happens, nothing can ever escape from the gravitational field of the star. The neutron star with such a tremendous density actually seals itself off from the rest of the universe. Things can get into the grip of the field, but they cannot get out. Such an anomaly is called a *black hole*. We will explore the mysteries of these bizarre theoretical objects in the next chapter. Some evidence has been found to suggest that such black holes actually exist.

INFRARED OBSERVATION

The radio frequencies have, as we have seen, much greater wavelengths than the visible or colored part of the electromagnetic spectrum. The shortest radio microwaves measure perhaps 1 millimeter or thereabouts in length; the reddest visible light has a wavelength of a little less than 1 micron, or 0.001 millimeter. That is a span of a thousandfold, and it is called the infrared range. In terms of frequency, the infrared lies below the red, and it is from this fact that it gets its name. Our bodies sense infrared radiation as heat. As you sit near an open fire and feel the hot glow on your face, you know that the fire gives off infrared radiation. Actually, the infrared rays are not themselves "heat"; but they produce heat when they strike an object, such as your body!

Stars, galaxies, planets, and other things in space radiate at all wavelengths, not just at those that are convenient for us to observe. In some portions of the infrared spectrum, the atmosphere of our planet is opaque. Between about 7500 angstroms (the visible red) and perhaps 2 microns, our atmosphere is reasonably clear, and it is possible to observe these wavelengths from ground-based locations. For longer infrared wavelengths, however, the observations must be made from high in the atmosphere, or from space.

The moon, the sun, and the planets have all been observed in the infrared, as have some stars and galaxies. Infrared observing equipment is very much like the optical telescope. Similar lenses and films are used, and excellent resolution can be obtained. Special kinds of film have been recently developed, making observation possible at longer and longer infrared wavelengths. Infrared astronomy has helped scientists to discover certain peculiar dim stars, which seem to radiate most of their energy in the infrared region. Visually, such stars appear red, and rather dim. But, like a faintly glowing electric-stove burner, they are powerful sources of infrared. It may seem paradoxical that these stars have relatively cool surface tempera-

tures compared with other stars; but the peak wavelength at which an object radiates is a direct function of the temperature. Cooler stars produce radiation at predominantly longer wavelengths than hotter stars. The hottest stars are intense sources of ultraviolet rays and X rays, and comparatively weak radiators of infrared.

Infrared astronomy is becoming more important in the study of evolving stars and star systems. As a cloud of dust and gas contracts, it begins to heat up, and its peak radiation wavelength becomes shorter and shorter. A cool, diffuse cloud radiates most of its energy in the radio part of the electromagnetic spectrum. Hot stars radiate largely in the ultraviolet and X-ray regions. At some point between the initial contraction of the cloud and the birth of the star, the peak wavelength must pass through the infrared.

The observation of infrared emissions is also important in the evaluation of dying stars. As a white dwarf cools down and becomes a black dwarf, its peak radiation wavelength decreases. On its way toward ultimate cold demise, the star must emit, at a certain time, most of its energy in the infrared.

The characteristic radiation of a star is a form of emission known as blackbody radiation. A blackbody is a theoretically perfect absorber and radiator of energy at all wavelengths. Any object having a temperature that is above absolute zero—about −273 degrees Celsius or −459 degree Fahrenheit—has a characteristic pattern of wavelength emissions that depends directly on the temperature. For any heat-radiating object (and that includes everything in the universe), the emission strength is maximum at a certain defined wavelength, and decreases at longer and shorter wavelengths. By observing an object at many different wavelengths, including the radio region, the infrared, the visible range, and the ultraviolet and X-ray spectra, the point of maximum emission can be found. Sometimes it can be inferred even if observations are not actually made at that wavelength. From the maximum-emission wavelength, the temperature of the object can be determined. Figure 3-10 shows the function that is employed by scientists for this purpose.

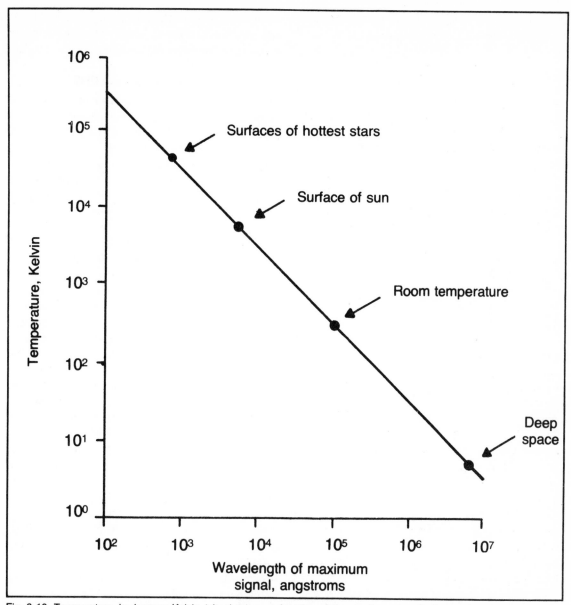

Fig. 3-10. Temperature, in degrees Kelvin (absolute), as a function of the maximum-amplitude wavelength of a body.

The capability of astronomers and physicists to make short trips into space will certainly be enhanced by the recent development of the space shuttle! Observations from space, at wavelengths that are blocked by our atmosphere, will help us to learn more about the mysteries of the cosmos. The far infrared, or the range of wavelengths longer than about 2 microns and extending into the upper part of the radio-frequency range, is not the only continuum of wavelengths at which the atmosphere presents an obstacle. The ionosphere absorbs or reflects much of the energy at the long radio wavelengths. Difficulty is also encountered at the shortest extremes: the ultraviolet, X-ray, and

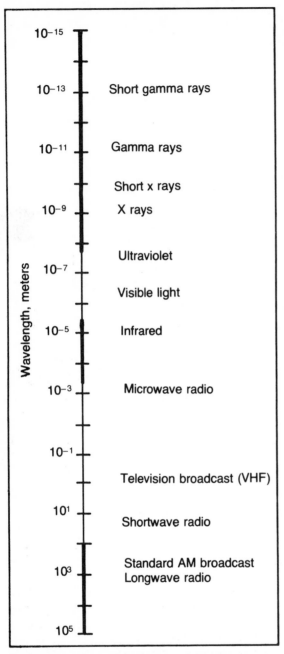

Wavelength, meters	
10^{-15}	
10^{-13}	Short gamma rays
10^{-11}	Gamma rays
	Short x rays
10^{-9}	X rays
	Ultraviolet
10^{-7}	
	Visible light
10^{-5}	Infrared
10^{-3}	Microwave radio
10^{-1}	
	Television broadcast (VHF)
10^{1}	Shortwave radio
10^{3}	Standard AM broadcast Longwave radio
10^{5}	

Fig. 3-11. The atmosphere is opaque at some electromagnetic wavelengths. The heavy lines in this illustration show the ranges in the electromagnetic spectrum at which the atmosphere is opaque; below about 100 meters, the ionosphere reflects incident radio signals back into space (lower extreme).

gamma-ray parts of the spectrum. Figure 3-11 illustrates the regions at which our atmosphere is opaque to electromagnetic waves.

ULTRAVIOLET ASTRONOMY

As the wavelength of an electromagnetic disturbance becomes shorter than the smallest we can see, the energy contained in each individual photon increases. The ultraviolet range of wavelengths starts at about 3900 angstroms and extends down to a length of perhaps 50 angstroms, although there is no precise definition of how short an ultraviolet wave can be. At any rate, however, the difference between 3900 angstroms and 50 angstroms is quite large. It is plainly a factor of several dozen.

At a wavelength of approximately 2900 angstroms, the atmosphere of the earth becomes highly absorptive, and at still shorter wavelengths than this, the air is essentially opaque. (This is a good thing, because it protects the environment against damaging ultraviolet radiation from the sun.) Actually, the atmosphere begins to scatter the radiation even in the visible blue part of the spectrum. This is why the sky appears blue to our eyes. Ground-based observatories can see something of space at wavelengths somewhat shorter than the visible violet, but when the wavelength gets down to 2900 angstroms, nothing more can be seen. At the shorter ultraviolet wavelengths, as in the case of the far infrared, it is necessary to put the observation apparatus into space.

Since glass is virtually opaque at ultraviolet, ordinary cameras with glass lenses cannot be used to take photographs in this part of the spectrum. Instead, a pinhole type device is used, and this severely limits the amount of energy that passes into the detector. While a camera lens may have a diameter of several inches, a pinhole is usually just a few thousandths of an inch across. This does not present too much of a problem for observations of the sun, which is very bright anyway; but for looking at things far away in space, it is not satisfactory. For the fainter celestial objects, the spectrophotometer is preferred. This device is similar to a spectroscope, in which a diffraction grating is

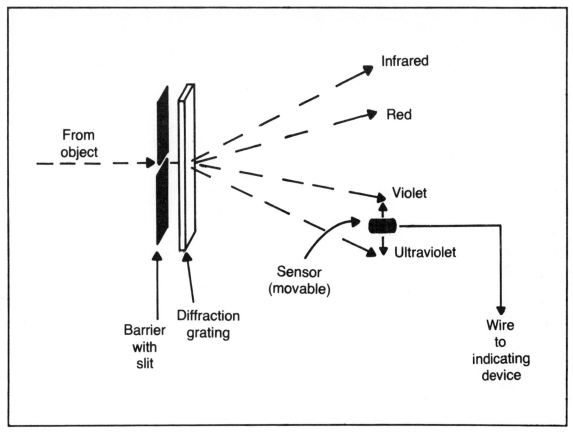

Fig. 3-12. A spectrophotometer employs a diffraction grating to split radiation into its constituent wavelengths. Then, a detector, movable back and forth along the spectrum, can be used to measure the intensity at specific wavelengths.

used to disperse electromagnetic energy into its constituent wavelengths. By moving the sensing device back and forth, any desired wavelength may be singled out for observation. The principle of operation of the spectrophotometer is shown in Fig. 3-12. At the extremely short end of the ultraviolet spectrum, radiation counters are sometimes used, similar to the apparatus employed for the measurement of X rays and gamma rays. For photographic purposes, ordinary camera film seems to work quite well at the longer ultraviolet; a special film, rather like X-ray film, is used at the shorter ranges.

The hot type O and B stars are strong sources of ultraviolet radiation. In fact, these stars evidently radiate more energy in the ultraviolet than in any other part of the electromagnetic spectrum. Type O and B stars are generally young stars. Within the visible part of the spectrum, their energy tends to be concentrated toward the violet and blue, and thus such stars look blue to us. The surface temperatures of these stars is much greater than the temperature at the surface of our sun. Type O and B stars have surface temperatures ranging from about 15,000 to 25,000 degrees absolute, or 27,000 to 45,000 degrees Fahrenheit; some are even hotter than this.

The type A and F stars radiate smaller amounts of ultraviolet energy than the O and B stars. Thus they do not seem as blue. Most type A and F stars look white. Still, these stars are hotter than our sun. Temperatures at their surfaces are in the range of about 8,000 to 15,000 degrees absolute, or 14,000 to 27,000 degrees Fahrenheit.

The sun is a type G2 star. Its surface temperature is in the neighborhood of 6,000 degrees absolute, or 11,000 degrees Fahrenheit. Much of the radiation from the sun occurs in the visible part of the electromagnetic spectrum. Comparatively little ultraviolet radiation is produced by our parent star. This is fortunate for the kind of life that has evolved on this planet; if the sun were a different type of star, certainly life would have developed in a different way. Some ultraviolet radiation is produced by the sun, however, and this radiation has been investigated using modern equipment aboard rockets and satellites. The findings have been very interesting.

The ultraviolet surface of the sun is somewhat above the visual surface. This tells us that, as the altitude above the photosphere increases, the temperature gets hotter. This fact also makes us wonder what, in fact, is meant by the expression "the surface of the sun"! If our eyes were responsive to a range of wavelengths just half as long as they actually are—say, a continuum of 2,000 to 4,000 angstroms instead of the actual 3,900 to 7,500 angstroms—the disk of the sun would look a little larger than it appears in the actual visible range. We would then ascribe to the sun a different photosphere. The sun would not only look slightly larger in size, but it would also look much less brilliant. The atmosphere of the earth is opaque in part of the range 2,000 to 4,000 angstroms. Assuming the longest detectable wavelengths appeared red to us, we would call the sun a decidedly red star!

Some creatures on our planet, notably certain insects, have eyes that are believed to be something like the hypothetical ultraviolet eyes we have just mentioned. If you could ask a fruit fly about the appearance of the sun and expect to get a meaningful reply, what would the fly tell you? The fly would describe a sun that subtends a slightly larger angular diameter than the sun you actually see. The fly would tell of a rather dim sun, and a very reddish sun. So, then, where is the surface of the sun, and what color is it? The only good answer to these questions are that the parameters of surface and color depend on the wavelength at which the sun is observed. This makes the sun a rather mysterious object. No planet changes in size with the wavelength!

The type K and M stars have the coolest surfaces of any stars, ranging as low as perhaps 1,500 degrees absolute, or about 2,700 degrees Fahrenheit. The greatest amount of visible radiation from such stars falls into the red and orange. These stars produce very little ultraviolet radiation.

Supernovae, or exploding stars, produce fantastic amounts of visible light, but even more of their energy output is in the ultraviolet range. The explosion of a supernova within a few light years of the solar system would certainly be a spectacular sight; the brilliance of the star would greatly exceed that of the full moon. But the ultraviolet radiation might be even more intense than that from our own sun. You could get a sunburn in the middle of the night, and certainly without realizing it, since the amount of heat would be relatively small. The shortest ultraviolet rays, which would cause the most damage to life on our planet, would be largely kept at bay by the atmosphere.

A supernova, after having thrown off a cloud of gas in the process of exploding, often causes ionization of the cloud because of ultraviolet radiation. Then, the glowing cloud can be seen from a great distance. Sometimes nearby stars, of the hot type O and B, cause ionization of interstellar gas. This produces such spectacular astronomical objects as the Great Nebula in Orion, the Horsehead Nebula, and others. Bright emission nebulae betray the presence of ultraviolet sources in their vicinity. A fluorescent light bulb operates on the same principle as the emission nebulae; the coating on the inside of such a bulb is set to glow by ultraviolet radiation from the gases within.

X-RAY ASTRONOMY

There is, as we have mentioned, no universally accepted cutoff wavelength at which the ultraviolet part of the spectrum ends and the X-ray region begins. Perhaps about 50 angstrom units is a

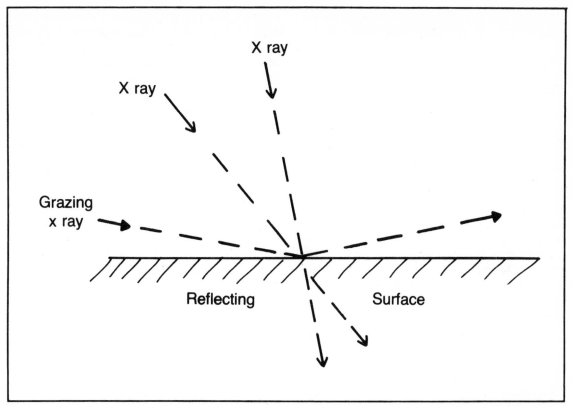

Fig. 3-13. When X rays strike a reflecting surface, they are absorbed unless the angle of incidence is sufficiently small.

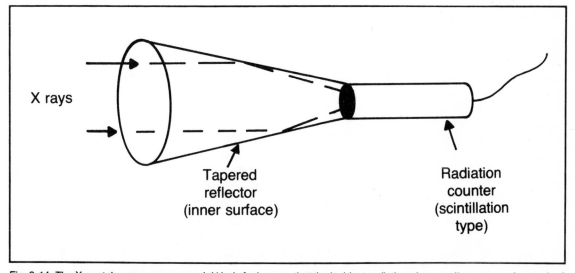

Fig. 3-14. The X-ray telescope uses a special kind of mirror, so that the incident radiation always strikes at a grazing angle. A scintillation detector is placed at the point where the rays come to a focus.

reasonable boundary. As the wavelength of energy gets shorter and shorter, it becomes more and more difficult to direct and focus the rays. This is because of the penetrating power of the short-wavelength rays. A piece of paper with a tiny hole might work very well for ultraviolet photography; but in the X-ray spectrum, the radiation will pass right through the paper as though it were transparent. If X rays land on a reflecting surface at a nearly grazing angle, however, and if the reflecting surface is made of suitable material, some degree of focusing can be realized. The shorter the wavelength of the electromagnetic energy, the smaller the angle of incidence must be if reflection is to take place: At the shortest X-ray wavelengths, the angle must be smaller than 1 degree. At still smaller wavelengths, and into the gamma-ray region, no reflection can be made to happen, no matter what the angle of incidence. This grazing-reflection effect is shown in Fig. 3-13.

A special cylindrical focusing mirror is used with the X-ray telescope. The mirror is actually not quite a perfect cylinder; it is tapered at an angle of a few degrees, and the taper is slightly nonlinear. Figure 3-14 is a rough illustration of how the X-ray telescope achieves its focusing. As parallel X rays from a distant celestial object enter the aperture of the reflector, they strike the inner surface at a very small angle of incidence. If the cylinder is tapered in just the right way, the X rays will be brought to a focal point. The radiation detector is placed at this point.

The resolution of an X-ray telescope, such as the one illustrated in Fig. 3-14, is not as good as that obtainable with optical apparatus, but it does allow the observation of some discrete celestrial X-ray sources. As is the case with ultraviolet radiation, X rays must be viewed from above the atmosphere of our planet. X-ray telescopes aboard rockets and satellites send their information back to the earth via radio transmitters, or record their findings on film and are returned to the surface for recovery.

After the development of high-altitude rockets and space vehicles, it became possible to look at the X-ray sky; this has only been within the last few decades. The curiosity of the astronomers was amply rewarded when the first results were obtained. Powerful X-ray sources were found, but there at first appeared to be no explanation for them; even the hot type O and B stars do not produce large amounts of radiation in the X-ray spectrum, since even they are not hot enough. Such stars do, of course, emit some X rays, but most of their energy is at longer wavelengths. The X-ray objects appeared to produce more energy in the X-ray region than at longer wavelengths.

Some tentative hypotheses have been brought forward to explain the X-ray objects, which have, because of their apparent location within our own galaxy, been called X-ray stars. They are not supernovae; their radiation wavelengths are too short even for that. They are not visually bright enough to be supernovae. And X-ray stars are much more commonly found than supernovae. But Margaret Burbridge, Herbert Friedman, and Alan Sandage have theorized that the X-ray objects are binary stars in very close mutual orbits. Matter from one of the stars in a binary system could be torn away from the other member by gravitational forces. The stars might even be in mutual contact. The exchange of matter between two stars in such close association could, these astronomers showed, account for the production of large amounts of X rays.

Another suggestion concerning the X-ray stars has been given: They could be binary star systems in which one member is a neutron star or very dense black dwarf. Still another theory holds that the strange stars are binary systems containing black holes. The gravitational influence of a neutron star or black hole would be sufficient to account for the X-rays; as matter was torn from the visible member of such a binary system, the hotter interior layers would be exposed, and this would produce radiation at very short wavelengths. The idea that matter is, in some way, being torn away from a star, is lent support by the existence of Doppler shifts in the X rays. But the mystery is far from resolved. Perhaps there are several different kinds of X-ray stars. X-ray astronomy is still in its infancy. There is much more to learn.

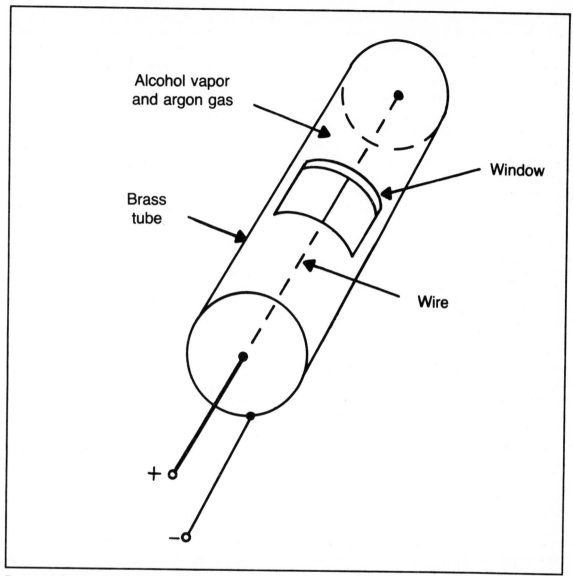

Alcohol vapor
and argon gas

Window

Brass
tube

Wire

+

−

Fig. 3-15. A Geiger counter contains a brass tube with a thin wire strung axially down the middle. The tube has a window, which may be opened or closed to allow less or more energetic radiation to enter. Inside the tube, a gas is ionized whenever a high-speed particle enters; this causes conduction for a brief moment between the terminals marked + and −.

Some X-ray objects seem to be outside of our galaxy; certain quasars and radio galaxies have been associated with strong sources of X rays. Iosif S. Shklovskii has hypothesized that interaction among the photons of radiant energy at different wavelengths, as they collide with each other, could be responsible for the X-ray emissions from ex-

tragalactic objects. It could be, also, that some galaxies and quasars have regions of tremendously hot material—hotter than anything we know in our Milky Way.

GAMMA RAYS AND COSMIC PARTICLES

As the wavelength of electromagnetic radiation becomes shorter and shorter, their penetrating power increases until focusing is impossible. The exact cutoff point where the X-ray region ends and the gamma-ray region begins is, as with the X-ray/ ultraviolet-ray boundary, not well established. But gamma rays can get shorter without limit. The gamma classification is the most energetic. Radiation counters are the primary means of observation at the gamma-ray wavelengths. The gamma rays from space are largely produced by high-speed atomic particles, and this kind of radiation therefore differs from the other electromagnetic energy, which is the result of photon bombardment.

Particles from space can be detected even at ground level by an ordinary radiation counter. The counter consists of a thin wire strung within a cylindrical metal tube, and certain gases are placed inside the sealed tube. When an atomic particle such as a high-speed proton, neutron, or atomic nucleus enters the chamber, the gas is ionized for a short time, and conduction occurs between the inner wire and the cylinder. A voltage is put between the wire and the outer cylinder, so that a pulse of current occurs whenever the gas is ionized. A simplified diagram of such a radiation counter is shown in Fig. 3-15. Matter, as you recall, is composed mostly of empty space, and this is true of the metal shell of the counter tube; thus high-speed atomic particles, which are very tiny yet massive for their size, have no trouble penetrating the tube casing if they are moving fast enough. The thicker the casing, the greater the speed the particles must have to get inside the tube. A window is usually provided in the tube, which may be opened to let in particles of lower energy, and closed to allow only the fastest particles to get inside.

Even while sitting in a room with no apparent radioactive materials, and with the window of the counter tube closed, there is an occasional click from a radiation-measuring device. Some of the particles come from the earth, since there are radioactive elements in the ground almost everywhere (though in very small quantities). Some of the radiation comes from space. The direction of arrival of high-speed atomic particles can

be determined, to a certain extent, by means of a cloud chamber. The air in a small enclosure can be treated especially to produce condensation when an atomic particle enters, and the path of the particle will then show up as a vapor trail.

It was in the early twentieth century that physicists noticed radiation apparently coming from space. They found that the strange background radiation increased in intensity when observations were made at high altitude; the radiation level decreased when observations were taken from underground or underwater. This space radiation has been called cosmic radiation.

The cosmic particles first discovered were what is called secondary radiation. The actual particles from space, called primary cosmic particles, usually do not penetrate far into the atmosphere before they collide with, and break up, the nuclei of atoms. To observe primary cosmic radiation, it is necessary to ascend to great heights, and as with the ultraviolet and X-ray investigations, this was not possible until the advent of the space rocket. Cosmic-particle astronomy is a very young science.

Cosmic radiation gives rise, in the upper atmosphere of the earth, to particles called mesons, and the mesons have provided an important vertification of the special theory of relativity. Fast-moving protons from outer space often break up the nuclei of air molecules, and in this process the mesons are generated. Mesons have very short lives before they degenerate, however; they last for only about a hundredth of a microsecond. We would not expect to find any mesons at ground level, then, because they cannot be expected to travel very far in that short time. Even at the speed of light, or 300 million meters per second, a meson would be able to travel only 3 meters. And yet, mesons are observed at ground level, and scientists can only explain it on the basis of relativistic time-space distortion. Because of their tremendous velocity, the distance from the ground to the top of the atmosphere is compressed. If we could ride on a meson, we would see the 100-mile distance as just a few inches!

Strange things happen to objects when they

travel close to the speed of light, as cosmic particles do. While the radiation in the electromagnetic spectrum—the radio waves, infrared, visible light, ultraviolet, X rays, and gamma rays—consists of photons traveling at the speed of light, the cosmic particles are actual particles of matter, traveling at speeds almost, but not quite, as fast as light. At such high speeds, the protons, neutrons, and other heavy particles gain mass because of relativistic effects, and this renders them almost immune to the effects of magnetic and electric fields in space. Such particles, arriving in the upper atmosphere, should come to us in a nearly perfect straight-line path, despite the magnetic field of our planet, and by carefully observing the trails of the particles in a cloud chamber aboard a space ship, it should be possible to accurately determine the direction from which they have come. In deep space, the direction of arrival would be ambiguous; a straight line points in two opposite directions. But in orbit around the earth, the ambiguity would be eliminated because one end of the line would point into the planet, and the other toward space. With the space shuttle now in operation, cosmic-particle astronomy will receive a great boost, because cloud chambers will be easily carried aloft and observed directly by physicists.

The potpourri of particles discovered by physicists, both by experimental and theoretical means, is mind-boggling. We have already observed neutrinos, those strange massless wisps that pass through planets and stars virtually unaffected. These particles are more fully discussed in Chapter 2. The neutrino experiments have taught us much about the sun and stars. But there are many other strange particles that have been predicted by theorists, based on their equations, and in some cases the particles have actually been found.

Anti-electrons, also called positrons, have been discovered; so have anti-protons and anti-neutrons. When an antimatter particle collides with its matter counterpart, they destroy each other! Theory has also predicted the existence of tachyons, which travel faster than light in apparent violation of the principles of relativity. Such parti-

cles have not been definitely found, but they should travel backwards in time and have imaginary mass! The list goes on, and will no doubt continue to grow; it is surely a surreal universe!

GRAVITY WAVES

According to the physics of Newton, gravity travels infinitely fast through space, but Einstein postulated that nothing could propagate faster than an electromagnetic field in vacuum. Einstein showed that gravitation creates distortion of space, and we will have more to say about this in the next chapter. Modern astrophysicists combined these two ideas of Einstein, and theorized that perhaps gravitational disturbances might be propagated through space in the form of waves. Apparatus was built to detect the so-called gravity waves, and they were indeed found!

You have no doubt observed how the impact of a pebble on a smooth water pond will cause ripples in the pond, which travel outward from the point of impact at a certain speed, and which have a certain amplitude and wavelength. It is a powerful mental exercise to imagine how gravity waves propagate through space, and how these waves affect material objects; but the water pond provides a good analogy.

Imagine a pond that is so calm that anything placed on its surface will remain in the same spot indefinitely. Suppose you place a small drop of dyed oil on the surface of the pond. The oil spreads out into a dot. Let the dot represent an object in the universe, dimensionally reduced so that we can imagine subsequent events. This oil dot may be a planet, a star, a stone, or anything you like—even your own body.

Now if you drop a small pebble into the pond near the oil dot, the disturbance of the impact will travel outwards in concentric circles. The surface of the pond represents the continuum of space-time, and the concentric waves represent the gravitational disturbance in space-time. What will happen when the waves reach the oil dot? Assuming the oil dot is large enough, some of the molecules of oil will be at a wave crest while others are in a wave trough. The shape of the oil dot, as measured over the surface of the water, will change as the waves undulate. Then, after the waves pass, the oil dot will ideally have its original shape. This is exactly what happens to an object in our universe when gravity waves pass, except that there is one more dimension involved: The object and the continuum are three-dimensional, and the wave disturbances are four-dimensional.

This idea, based on Einstein's special and general theories of relativity, led scientists to look for gravity waves. The detecting apparatus consisted of a long metal cylinder, with sensitive devices attached. The size and shape of the cylinder were very accurately ascertainable; the equipment was so sensitive that local effects, such as people walking around the device, actually registered! It was necessary, then, to use two identical gravity-wave detectors, separated by a great enough distance so that terrestrial effects could be ignored. The two instruments did occasionally register simultaneously; this indicated the arrival of gravity waves from space!

What might generate gravitational waves? The space-time continuum of our universe is apparently criss-crossed with such disturbances; the cosmos is not like a calm pond, but instead resembles a windswept ocean. Collapsing stars, sufficiently heavy to affect space and time to an infinite extent, are believed to be one source of gravitational waves. If a star has enough mass as it cools off and contracts, it will literally withdraw from our universe as its gravitational field becomes so powerful that nothing, not even electromagnetic waves, can escape its grasp. At the precise moment this happens, and the star becomes what is called a black hole, gravity waves could be sent out in a burst.

Then the star would be gone forever!

Many astronomers believe that massive black holes, consisting of collapsed congregations of stellar matter, exist at the centers of galaxies. Such black holes would continually pull stars in from the outside, and if this is happening, then we ought to find gravity-wave bursts coming from the centers of galaxies. High resolution is not yet possible with gravity-wave receiving equipment, but the direction from which a disturbance arrives can be determined to some extent. It seems quite possible that many, if not most, galaxies have black holes at their centers.

The very sound of it—a black hole—provokes thoughts of something strange. Black holes certainly are strange; they are perhaps the most incredible things in the cosmos. A black hole is a knot, a twist, in the continuum. Time is either accelerated into the infinite hereafter, or else brought to a complete stop, depending on the point of view from which the hole is seen. We have never made a black hole in a laboratory, and no black hole has ever smashed into our planet or our sun. Black holes were discovered, as things in the physical sciences often are, by theoreticians working with mathematical equations. This allows the disbelievers a way out, since they can always say, "Show me one." If that could be done, however, they would not survive the encounter.

The concept of the black hole arose from Einstein's general theory of relativity. This theory demonstrates that the universe can be defined from any point of view. Mathematically, this involves the stretching and bending of time and space. Let us look at the basic idea of general relativity. Then we will see what happens when gravity is allowed to run wild!

Plate 1. The center of our Milky Way galaxy lies in the direction of the constellation Sagittarius, the archer. This is where most of the stars are concentrated. Interstellar dust clouds obscure much of our view of the galactic center. If there were no such clouds, we would have almost 24 hours of daylight for much of the year. (U.S. Naval Observatory photograph.)

Plate 2. The Veil Nebula in Cygnus is a rapidly expanding cloud of glowing gas. The outward movement near the edges of this nebula has been measured by Doppershift methods, and is about 250,000 miles per hour, indicating that the nebula is the result of a supernova explosion. Many thousands of years ago, the Veil Nebula probably looked like the Crab Nebula in Taurus. (U.S. Naval Observatory photograph.)

Plate 3. The Trifid Nebula in Sagittarius gets its name from the three dark filaments that extend outward from its center. The Trifid nebula is the birthplace of new stars. The ultraviolet radiation from the hot young stars causes the gas to fluoresce. Our own sun was originally part of such an interstellar gas cloud. The Trifid Nebula is 3,500 light years away and about 15 light years in diameter. (U.S. Naval Observatory photograph.)

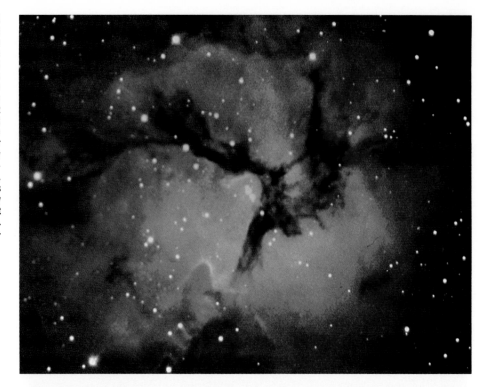

Plate 4. Another site of star formation is the Lagoon Nebula, also in Sagittarius. The dark flecks in this gas cloud are protostars, or regions in which the hydrogen is undergoing gravitational collapse. Eventually, these condensed areas will become so dense that the fusion reaction begins. Hundreds of new stars will form from this nebula. Some young stars have already begun to shine in the center of the cloud, causing fluorescence. (U.S. Naval Observatory photograph.)

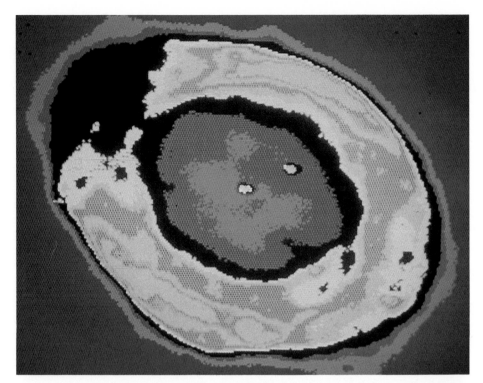

Plate 5. An electronic photograph of the Ring Nebula shows internal details not visible in most ordinary photographs. The colors are artificially produced by a charge-coupled-device camera. The Ring Nebula is the site of a violently exploding central star. The expanding cloud of gas will eventually disperse into space, and new stars will form from it. (Smithsonian Institution, Multiple Mirror Telescope Observatory.)

Plate 6. This interstellar cloud has been called the Dumbbell Nebula because of its physical shape. An aging star at the center of the nebula causes the gas to glow. Astronomers of the early nineteenth century, seeing nebulae such as this through their small telescopes, mistook them for planets because of their large angular diameters. Today, for this reason, we call these glowing clouds planetary nebulae. (U.S. Naval Observatory photograph.)

Plate 7. This spiral galaxy in Coma Berenices has a large amount of dark material in its tightly coiled arms. We view this galaxy from well above the plane of its equator. This is a type Sb galaxy. Some spirals have arms that are even more tightly coiled. (U.S. Naval Observatory photograph.)

Plate 8. One of the most beautiful galaxies is the so-called Whirlpool spiral in Canes Venatici. This galaxy is a type Sc spiral, with loosely coiled arms, and an external irregular component. The Whirlpool is about 14 million light years away from us. The bright nucleus is a dense congregation of relatively cool stars. Young stars and protostars are found in the arms of the spiral. (U.S. Naval Observatory photograph.)

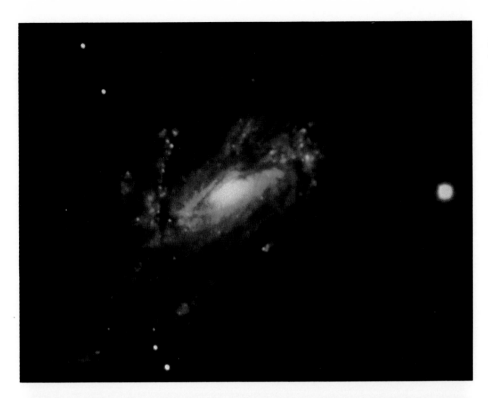

Plate 9. This type Sb spiral galaxy, found in the constellation Leo the lion, appears surrounded by wisps of stray stars above and below the equatorial plane. The outer regions of this spiral contain gas clouds from which new stars are forming. As with virtually all spirals, the bright core is made up primarily of older stars. (U.S. Naval Observatory photograph.)

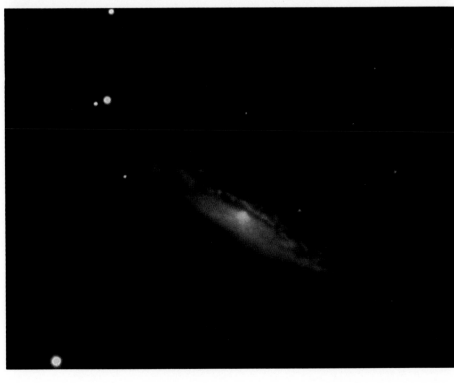

Plate 10. This spiral galaxy, in the constellation Lynx, presents itself nearly edgewise to us. From this angle, the dark nebulae are clearly visible. Most of this material lies within the plane of the galaxy, as do the stars. The dark clouds slowly rotate around the center of the galaxy like a vast cosmic hurricane. The motion produces smaller eddies, which eventually become star systems. (U.S. Naval Observatory photograph.)

Plate 11. Spiral galaxies may appear almost disk-shaped. The arms of this galaxy in Canes Venatici are difficult to see, except at the outer edges. Galaxies take on their own individual appearances; like snowflakes, no two are exactly alike. But each galaxy contains millions or billions of stars; most spirals are comparable in size to our Milky Way. (U.S. Naval Observatory photograph.)

Plate 12. These small elliptical galaxies are actually satellites of the Great Nebula in Andromeda, about 2 million light years from us. Most galaxies in the universe are elliptical, without identifiable spiral arms. Some elliptical galaxies are almost perfectly spherical, while others are more elongated than a football. Ellipticals vary in size from small satellites, such as these, to massive congregations of trillions of stars. The largest elliptical galaxies have about 100 times as many stars as our Milky Way. (U.S. Naval Observatory photograph.)

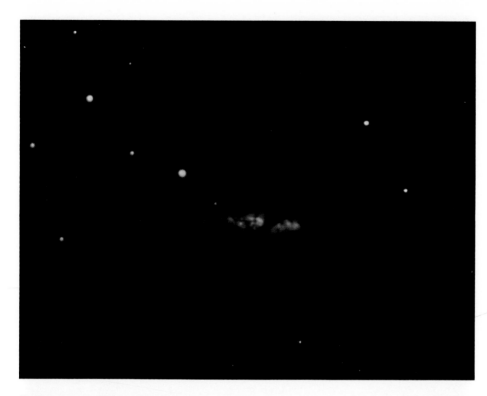

Plate 13. This galaxy, in the constellation Ursa Major, confuses astronomers who attempt to categorize it. Some think it is a disorganized spiral seen edgewise. Others believe it is an irregular galaxy. Because of the disagreement, this galaxy has been called a peculiar galaxy. Although this group of stars is smaller than our Milky Way, it radiates more energy. The reason for the unusual radiation intensity is not known with certainty, but it is possible that an explosion is taking place near the center. (U.S. Naval Observatory photograph.)

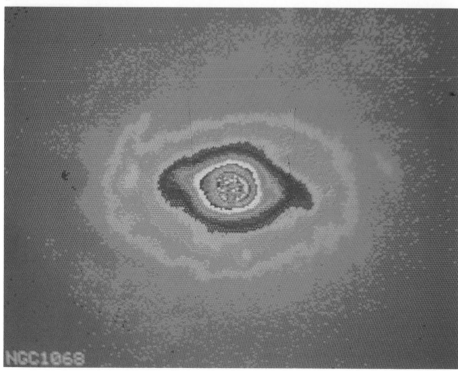

Plate 14. This is an electronic photograph of galaxy NGC 1068. The colors are artificially produced to show different brightness regions within the galaxy. Some galaxies are believed to harbor black holes at their centers. Using graphic renditions such as this, researchers hope to uncover some of the secrets of galactic interiors. (Smithsonian Institution, Multiple Mirror Telescope Observatory.)

Plate 15. The galaxy NGC 2787, electronically enhanced with artificial color, shows a more uniform structure than NGC 1068. The colors are chosen by the astronomers for various research purposes. This is a 5-minute exposure with the 1.5-meter telescope at the Fred Lawrence Whipple Observatory. (Smithsonian Institution, Whipple Observatory.)

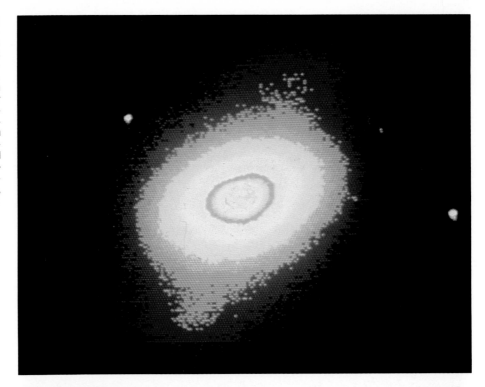

Plate 16. A more detailed view of the central structure of galaxy NGC 2787. The astronomer chose to use different colors, and a different scale, in this rendition, for the purpose of probing deeper into the nucleus. Several concentric regions of increasing energy output, not visible in Plate 15, can be seen here. (Smithsonian Institution, Whipple Observatory.)

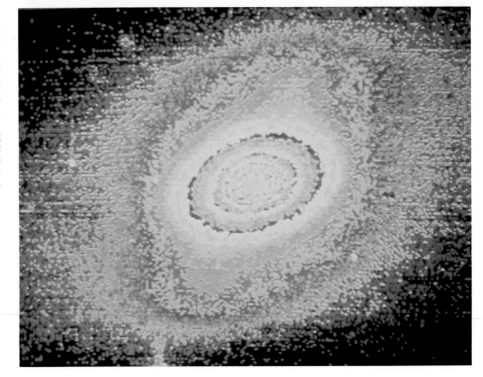

Chapter 4

The Black Hole: A Twist in Time and Space

WE HAVE ALWAYS BEEN INQUISITIVE CREATURES. The unknown holds a certain kind of attraction for our minds. We are, by nature, explorers. When exploration becomes impractical or impossible because of physical limitations, we turn to nonphysical vehicles such as mathematics and theoretical physics to satisfy our need for continual discovery. Intuitive universes, while perhaps less "real" than the physical cosmos, offer respite from the nagging imperfections of our existence. No one knows this better than the pure mathematician, whose universes can take on forms limited only by the boundaries of imagination. You cannot touch a number, or a variable, or a geometric point, plane, or line; but you can imagine such things without difficulty. These are purely theoretical concepts. On the other end of the intuitive spectrum, we find the experimentalists, whose realm is solely comprised of things that can be touched, seen, heard, and measured.

Somewhere in between the purely theoretical and the wholly physical, we will find the black hole! Black holes seem to be oddly perfect—too perfect,

perhaps, to be real things. But astronomers have every reason to believe that black holes are very real, and not only that, they carry their own reality with them into another universe, and another time! In order to get some understanding of how this can happen, without getting into an involved mathematical jungle, it is first necessary to have a basic idea of what led up to the idea that there might actually be such things as black holes.

THE EARLIEST CONCEPT

The black hole is not a twentieth-century invention, although it is the modern scientist who has refined the concept and given it the attention it deserves. John Michell, an eighteenth-century astronomer and mathematician from England, realized that the escape velocity from a celestial object depends on its mass and density, and that this escape velocity might become greater than the speed of light. Escape velocity is the minimum speed that an object must be given if the object is to get away from the gravitational influence of a planet or star. How hard would you have to throw a baseball, for example, if you wished to throw it out of the gravitational field of our planet? The answer is 25,000 miles an hour; obviously, no human arm is strong enough to accomplish such a feat. If the planet were more dense or more massive, or both, so that the gravitational field was stronger, the escape velocity would be greater. John Michell knew that the speed of light, while very great, is nonetheless finite. Any finite quantity can be exceeded, regardless of how large it is. If the escape velocity were to exceed the speed of light, Mitchell reasoned, then a star or planet would disappear from view. It would look black.

Michell did not receive a very enthusiastic response from the general scientific establishment. The reaction was something like, "So what? Show us one." The principles of relativity theory were not yet well refined, and the full extent of the implications of the black hole could not be understood. Curvature of space and time were unknown concepts. The art of observational astronomy was not sufficiently developed to allow good evidence to be found for the existence of black holes. With modern techniques for observing the cosmos at all wavelengths in the electromagnetic spectrum, today's astronomers think that they have detected black holes in space. But the final test is yet to come. We must find a black hole and enter it. A final test it will be, for if our theoreticians are correct, we will, upon entering a black hole, never see our universe again!

Before we let our imaginations dive into a black hole, let us first understand a little about the theories that led to the concept. It began with Albert Einstein's special and general theories of relativity. Einstein believed that all points of view in our universe must be equivalent; the idea of favored and less favored reference frames was objectionable to him.. This led first to the special theory of relativity, in which Einstein explained how all points of view are equivalent as long as they are intertial, or nonaccelerating. The general theory of relativity, perfected a few years later, finally provided the all-encompassing unity so much sought after: The universe can be regarded from any point of view whatsoever, and can always be considered to have the same properties.

After this great theoretical triumph of physics, it appears that discontinuities remain. They are other universes. They are black holes.

THE WAY YOU SEE IT

Our senses are our link with the world. We act according to how we see, hear, feel, taste, and smell. Reality has no other immediate effect on us, aside from how we sense it. But the way in which we sense things—and in particular, see them—depends on our point of view. Einstein realized this when he postulated that the speed of light, and of all electromagnetic radiation, does not depend on the motion of an observer, but instead, always seems to be the same. This bold axiom eventually led to the theory of relativity, and all its implications for modern science.

Imagine, for a moment, how difficult it is to measure something from a great distance. Suppose your friend holds a rod up in the air, while he stands several hundred feet away from you. If you know the distance to the rod, but you do not know the actual length of the rod, you must know the orientation of the rod before you can use trigonometry to figure out how long it is. If the rod is oriented perpendicular to your line of sight, it might subtend a certain angle in your field of vision; the more inclined the rod to your line of sight, the shorter it will seem to be. At a great distance, even through a telescope, you cannot readily tell if the rod is oriented perpendicular to your line of sight. So the length of the rod, as you see it, can vary anywhere from zero to its actual length! By its "actual" length, we really mean its length as your friend sees it. If he observes the rod as 10 inches long, but holds it at a 30-degree angle to your line of sight, you will think it is only 5 inches long. If he holds the rod right along your line of sight, you might think he has no rod in his hand at all, but instead is just raising a fist in the air. If you could not communicate with him, and could not go up to him and find out his point of view, then your "reality" could only be as you saw it. A rod has a length only if you can measure it.

According to the theory of relativity, the speed of light does not depend on the point of view of an observer; it does not behave like the rod in the imaginary experiment just preceding. But time, distance, and mass do behave something like the rod; our perception of these variables depends on our point of view. We measure the light from the stars as arriving at about 186,282 miles per second. We see the light from the moon, or the sun, or a desk lamp as arriving at this same speed. If we were to board a space vessel and move toward or away from the sun, we would find that the speed of the arriving light would not change, no matter what our velocity! This idea is intuitively difficult for some people to accept. It would seem that, if we were moving away from the sun at 1,000 miles per second, then we would see its light coming toward us at 185,282 miles per second. Einstein saw things differently. His axiom has since been verified using accurate measuring apparatus and moving vehicles.

If we accept Einstein's axiom of the constancy of the speed of light—and we must, since it has been demonstrated to be true—then high speeds have certain effects on time, space, and mass. Isaac Newton thought that time flows smoothly, and that its rate does not depend on anything except the substance of the universe. But the rate of time is as we measure it. From two reference frames that are moving with respect to each other, time will seem to flow at different rates.

MOTION AND THE FLOW OF TIME

In order to measure time, we must build a clock. In order to have a clock, we must have a reference interval, such as the period of moon around the earth, or the period of revolution of the earth around the sun. We assume that the reference interval is itself constant, in terms of elapsed time. If the reference interval is not uniform, then we cannot make any arrogant claims about the validity of our time measurements!

The speed of light, since it has been postulated to be independent of the point of view of an observer, is perhaps the best reference we can use for time measurement. Suppose we built a space ship, having a diameter of 983 feet, 7 inches. You can

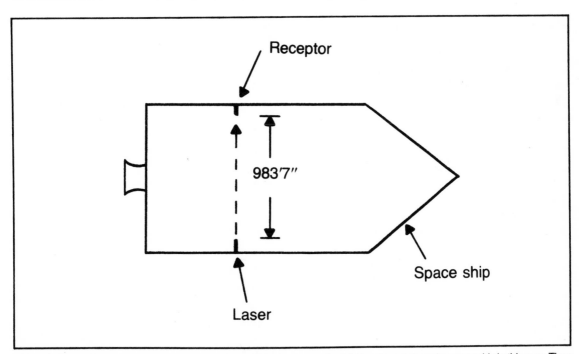

Fig. 4-1. A light-beam clock, having a unit time length of 1 microsecond, might be placed aboard a space ship in this way. Time intervals can then be measured when the ship is moving at various speeds.

verify for yourself that this is 0.186282 miles, and since light travels 186,282 miles per second, a laser beam would take exactly a millionth of a second to get across this space ship. Figure 4-1 shows this apparatus for measuring time; it is aboard a space ship because we will, in our minds, need a mobile clock in the cosmos. The fundamental unit of time will be, for us, 1 microsecond, or a millionth of a second: the length of time it takes a laser beam to cross the ship exactly sideways.

If we get into our ship and get it going, the light beam will always seem to travel directly across the inside of the vessel, except when we accelerate. (For now, we will have to put off discussion of what happens when we accelerate.) No matter what our speed, the light beam will always seem to follow the same path across the ship. We will therefore continue to rely on our laser clock for time measurements. We could bring along some other kind of high-speed clock to check the time required for the laser beam to travel across the ship, and we would always find the crossing to take 1 microsecond.

But suppose we were to watch this situation from the outside of the ship. This might be done by having the ship transmit a radio signal as the laser beam is first emitted, and another radio signal when it arrives at the other side of the ship. The time interval would appear to be slightly longer than 1 microsecond if the ship were moving relative to us. Figure 4-2 shows why this would be. To get from one side of the ship to the other, the beam of light would have to travel a slightly diagonal path. The path would still be straight, but not exactly sideways to the ship: The ship would move forward somewhat while the beam was crossing. The greater the speed of the ship relative to us, the sharper the angle at which the light beam would seem to cross, and the farther it would appear to travel in crossing the ship. Our own laser clock,

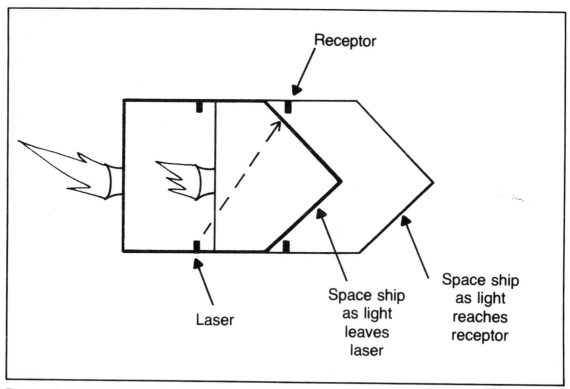

Fig. 4-2. When the light-beam clock is in motion, the time interval becomes longer because the light beam must travel a greater distance.

having the same physical dimensions as the clock on the ship, would give a different indication. This effect is not just a mathematical construct invented to stimulate the mind; it actually happens. Time does progress at a rate that depends on relative velocity. The phenomenon has been verified by scientists using extremely precise atomic clocks aboard moving airplanes.

As the speed of a moving object approaches the speed of light, the rate of time flow gets slower and slower as seen from an external viewpoint. At speeds near that of light, time almost stops; and at the speed of light, time actually does come to a halt, according to the equations. This has been called time dilation. But time is not the only parameter of the universe that is affected by motion. Spatial distance is compressed, and mass is increased.

MOTION AND SPACE

As an object moves with respect to us, its length in the direction of motion—that is, along the axis of its motion—becomes shorter. This happens only to a tiny extent until the speed becomes an appreciable fraction of the speed of light. The factor by which an object gets physically shorter is the same factor by which time moves more slowly; if v is the speed and c is the speed of light, then the distortion factor is equal to

$$\frac{1}{\sqrt{1-v^2/c^2}}$$

If v is small compared to c, then the expression is just about equal to 1. But when v becomes a sizable fraction of c, the expression gets larger and larger without bound.

The distortion of space caused by high speeds is somewhat analogous to the viewing of a rod from different angles, except that it happens in four dimensions instead of in three dimensions. We see things not just in space, but in time-space.

The shortening of distances at high speeds accounts for the observation of mesons on the ground, when we should not expect to find them. The mesons are tiny, subatomic particles, having a life span of about a hundredth of a microsecond. They are generated in the upper atmosphere of our planet as cosmic particles strike the atoms of the air. In a hundredth of a microsecond, we would expect the mesons to travel only 3 meters, not nearly far enough to reach the ground; but because of their great velocity, the distance from the upper atmosphere to the ground seems shortened by a large factor. Instead of tens or hundreds of miles thick, our atmosphere appears, to a meson, to be only a few inches deep! The existence of the mesons at sea level has been taken as proof that the spatial distortion, predicted by Einstein, actually does take place.

MOTION AND MASS

The famous relativistic time-dilation expression,

$$\frac{1}{\sqrt{1-v^2/c^2}}.$$

, applies not only to time and length, but also to mass. The relativistic effect on mass is of much more practical concern for future space travelers than the effect on physical length. For, as a space ship is brought near the speed of light, it gets more and more massive without limit. This mass increase makes it harder and harder to accelerate to greater speeds, because of the increased inertia. It is impossible to reach the speed of light using a finite amount of energy, because of this increase in mass with speed.

At high speeds, a space vessel itself would become more massive with respect to the overall cosmos, but the passengers would not notice this change. Nor would they notice the change in the rate of time progression and the distortion of their physical dimensions. But if a space ship moves at near-light speed with respect to the subatomic particles in interstellar space, the masses of those particles will appear greatly increased. Large amounts of energy will be released when a proton, neutron, or alpha particle collides with the space ship. The resulting radiation would prove quite deadly unless some means were devised to protect the occupants of the vessel. Besides the radiation hazard, tiny meteorites would present a great danger, as well. A small pebble, with a weight of an ounce, would grow in stature to a weight of several pounds, and its density would thus get much greater! At a high enough speed, the pebble might go through the ship.

The effects of high speeds on interstellar and intergalactic travelers will be hard to deal with, should our descendants ever develop the technology to attain relativistic speeds. But relative motion is not the only thing that affects the rate of time flow and the mass and shape of an object in space. Acceleration and gravitation have similar effects, and it is these effects, in particular, that create the black hole. The relativistic effects on time, space, and mass occur because of distortion in space itself. Space can be stretched, squashed, bent, and twisted. In order to understand this, we must first think a little bit about how we locate objects in space. Distortion in space will affect where, and how, we see an object from different points of view.

COORDINATE SYSTEMS

We define the positions of all objects by means of coordinate systems. On an elementary, subconscious level, you do this all the time. When you get up in the morning, you find your way to the bathroom to brush your teeth; how do you know where the bathroom is? In your brain, millions upon millions of tiny cells are cogitating, telling your body to move over x units, forward y units, and so on. You do not set up a rigorous mathematical system by conscious effort, but you could do so, and you could define the position of your toothbrush by a set of coordinate values. To uniquely and precisely do this, you would need three coordinates; we might call them x, y, and z.

Probably the most familiar coordinate system in three dimensions is the Cartesian coordinate system. This system employs three number lines, graduated in units of our choice, perhaps inches or feet. One number line runs vertically, and the other two horizontally, in such a way that all three lines intersect at a single, common point. All three lines are mutually perpendicular at that intersection

point. We can specify the dimensions of any object in terms of its height, width, and depth. Figure 4-3 illustrates an example of the three-dimensional Cartesian coordinate system.

Ordinarily, in a coordinate system for determining position, we insist that all three number lines have units of the same size. We insist that the lines be mutually perpendicular at the origin or point of intersection. This just stands to reason; things are made simple that way. We might even tape strings to the walls, floor, and ceiling of a room to illustrate the x, y, and z coordinate axes, knotting the strings together in the exact center of the room. (Try this; it's fun!) Using a felt marker, graduations can be put on each string. Then, any point in the room can be defined by one set of coordinates (x,y,z). Any set of values (x,y,z) corresponds to exactly one point in the room. Of course, if we choose numbers that are too big, the point might lie outside the room, but we can mentally extend the three straight axes indefinitely outward, and by so doing, we can define points in other rooms, in other neighborhoods, in other cities—even in other galaxies, if we are so motivated!

So far, we have been considering the appearance of this coordinate system only from a point of view not moving or accelerating with respect to the origin. We have not considered what this system might look like if our own reference point is changing. If we move at a slow speed in the system, it will look just the same as it does when we stand still. If you walk around the room (ducking under the strings when necessary), the coordinate system will look different from various angles, but the three axes will remain perpendicular to each other, and the graduations will stay the same size. But, suppose you were to walk across the room at relativistic speed, so that the time-dilation factor would influence your perception of the universe. Then the three axes would not necessarily seem perpendicular; the graduations on the axes would not necessarily all be the same size. The greater your speed, the more the axes would appear to change orientation, and the greater the difference in their unit lengths would become.

The above example is an illustration of how

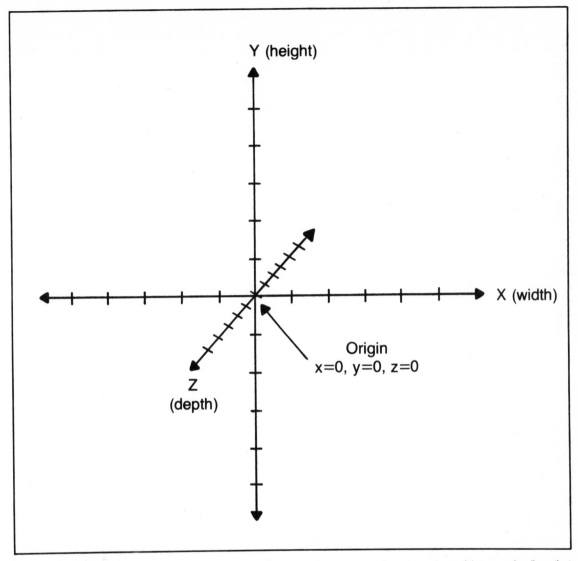

Fig. 4-3. The three-dimensional Cartesian coordinate system, in its most common form, is made up of three number lines that intersect at right angles.

space becomes distorted according to the special theory of relativity. At high speeds, the "shape" of space gets changed. Space becomes, so to speak, squashed or stretched. The coordinate system would no longer look the same to us, if we were moving at a fast enough speed. But it would still be a perfectly legitimate coordinate system. We would still be able to define the position of any point in space. The old coordinate system (as seen at rest)

and the new one (as seen from a moving point of view) would be related by what we call a linear transformation. This is just a mathematical way of saying that the strings would always look straight, no matter how fast our speed. Their intersection angle might change, and the size of the graduations might become nonuniform, but the strings would still appear straight.

If were to accelerate, or change velocity, at a

great enough rate, then the strings would appear to become curved. This might lead us to believe that space itself had become curved!

GEODETIC LINES

For a moment, let us get back to the stationary point of view, from which the axes all seem perpendicular, and from which the graduations all look the same. What makes us believe that the coordinate lines are straight? How might we verify this?

A simple way, and in fact the only real way, to ascertain "straightness" of a line is to use light beams. Light beams always travel in straight lines, do they not? If we were to check the strings in the room, using a laser beam, the strings would appear straight, assuming we had stretched them very tight! From the high-speed point of view, the strings would also look straight. But if we were to accelerate, or change velocity, at a rapid rate, then the strings would look curved. But the path of a laser beam would also seem curved, and this might make us wonder what a straight line really is. Actually, light does not necessarily travel in straight lines. Instead, light follows what is called a geodetic path through space.

We have all heard the statement that the shortest path between two points is a straight line. But actually, the shortest path between two points is a geodetic line, or the path that a light beam would follow to get from one point to the other. From some points of view, a geodetic line looks straight, but from other reference frames, it does not.

Acceleration is a change in velocity, and velocity is defined in terms of speed and direction. Acceleration can be manifested as a change in speed, such as when you hit the brake pedal in your car. Acceleration can also take the form of a change in direction, such as when you go around a curve. Acceleration can be a combination of changes in speed and direction. How do you tell whether you are accelerating or not? You always know that you are accelerating when you feel force. When you hit the brakes, you are pushed forward in your car; when you turn a corner, you are pushed outward.

For acceleration to significantly affect the path of a beam of light, the rate of velocity change must be extreme. Figure 4-4 shows one example of the effect of acceleration on a beam of light, as seen from two points of view.

Space travelers are subjected to intense acceleration, and this can have great effects on their bodies. To observe the effects of intense acceleration, researchers have routinely used a device called a centrifuge. The centrifuge consists of a chamber at the end of a mechanical arm in a circular room. The arm swings around and around, and the person in the chamber is subjected to centrifugal force that depends on the speed of revolution. Suppose you were to get inside a centrifuge and allow yourself to be subjected to extreme acceleration. Suppose that the speed of the chamber around the center of the room were to be increased until it was a sizable fraction of the speed of light! (If this were so, you would be killed by the force, but let's ignore that.) Imagine that you are equipped with a laser device in your little chamber, with which you can establish geodetic lines.

First, let us take the point of view of an observer outside the chamber, looking in at you as you perform your experiments (Fig. 4-4A). From this point of view, light beams always appear to travel in straight lines; this is true no matter where the light beam originates. The outside observer, watching you whirl around and around, would see the light from your laser moving in a straight line, just as he would see the light from a laser of his own. But if your speed of revolution were great enough, the centrifuge chamber would travel a few degrees of arc around its circular path even as light would traverse the distance across the chamber!

Thus, although you might aim the laser beam at a screen directly across the chamber, the beam would miss the screen entirely if the rotational speed of the centrifuge were great enough. In Fig. 4-4, the centrifuge goes around so fast that the light beam hits the floor. From the external point of view (at A), it is obvious that this must happen; the motion of the chamber literally carries it away from the beam. But as seen from inside the chamber, the path of the beam seems curved (as shown at B). From your point of view inside the chamber, you

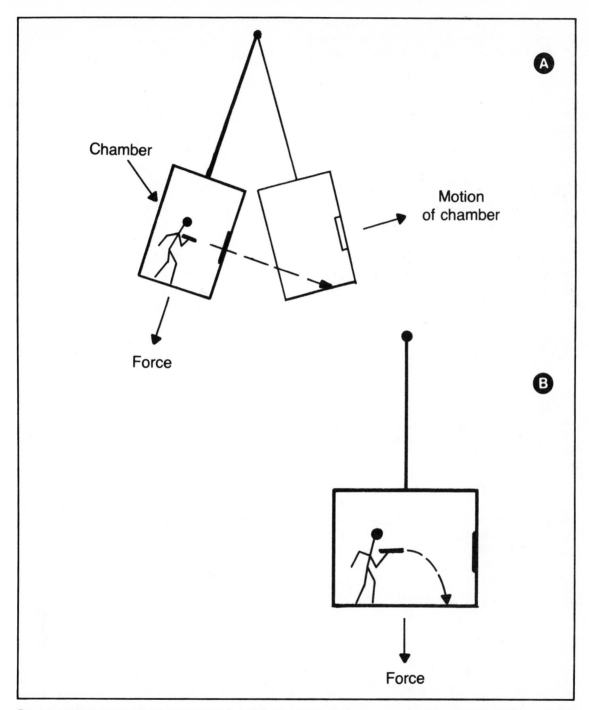

Fig. 4-4. A light beam seems to travel straight, from an outside point of view, as a centrifuge spins (A). The light beam misses the screen because of the motion of the centrifuge. As seen from inside (B), the light beam seems curved, and a passenger inside the chamber sees curved space.

would not sense the rotation of the chamber. You would know that there was a powerful force pulling at you, and you would assume that it was responsible for the curvature of the path of the light beam.

This example illustrates how geodetic lines in some universes can be curved. In this experiment, if it could be carried out, the laser beam would follow a geodetic path, no matter what the point of view. The greater the acceleration, the greater the distortion of the light beam. As the rotational speed of the centrifuge chamber approaches the speed of light, the distortion of the laser path becomes greater and greater. The difference between the "shapes" of the universes, one inside the chamber and the other outside, grows. If the chamber could somehow be flung around at the speed of light, the two universes would become entirely invisible with respect to each other.

TWO-SPACE

It is a difficult exercise to envision the curvature of three-dimensional space, and for this reason, it is convenient to imagine a two-dimensional continuum. Such a universe has no thickness at all. A two-space might pass through your body without exerting any influence whatsoever on you! But let us imagine that we could, in fact, see such universes embedded in ours. They would look like thin membranes of various shapes and sizes.

There is no reason why a two-space, consisting of a set of points in our three space universe, would have to be flat. (As a matter of fact, a perfectly flat two-space would be an oddity; curvature can be manifested in an infinite number of different ways, but flatness can happen in only one way!) The two-space might resemble the surface of a balloon, and be closed upon itself. It might have any geometric shape you can think of. How would the inhabitants of a curved two-space know it was curved, and not flat? To them, the idea of curvature in their universe might seem as incredible as the concept of curved three-space seems to us. But there are certain ways to verify that a continuum is curved or, as the mathematician would say, non-Euclidean.

Imagine a two-dimensional continuum that is curved locally as shown in Fig. 4-5. The creatures

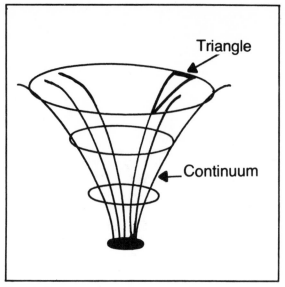

Fig. 4-5. Space curvature is negative, or funnel-shaped, near a source of gravitation. The sum of the measures of the angles in a triangle is, in such space, less than 180 degrees.

inhabiting the two-space universe could detect this curvature in a number of ways. One method would be to set up three space stations, and direct lasers among them so as to form a geometric triangle. No doubt the creatures would have already perfected geometry, and would know that the sum of the measures of the interior angles ought to be 180 degrees. But the experiment would not yield that result; the sum of the angles would be smaller than 180 degrees. The curvature shown in Fig. 4-5 is representative of the kind of distortion that cosmologists believe takes place in our universe, near any source of gravitation. The greater the intensity of the gravitational field, the most severe the distortion.

Imagine a two-dimensional universe that is shaped like a sphere, as shown in Fig. 4-6. If the triangle experiment were conducted in this universe, with a great enough separation among the space stations, the sum of the interior angles of a triangle would exceed 180 degrees. Although cosmologists believe that our universe is shaped like a four-dimensional sphere with a three-dimensional surface, the estimated circumference of the sphere is overwhelmingly large, and it will probably never be possible for us to conduct the laser experiment illustrated by Fig. 4-6.

Another way to detect the curvature of space is by the measurement of distances. Distortion of space makes things literally get farther away! If the continuum shown in Fig. 4-5 were flat, and not curved, two space stations would be closer together. Distance has meaning only if it is measured. The beings in the two-space would measure distance within their continuum; to us, their goedetic lines would not represent the shortest possible paths among points. The greater the distortion, the greater the distance would become. If the distortion of space were to get great enough, the distance would increase to many times its value as measured in a flat universe. In the black hole, space becomes infinitely distorted, and distances became greater and greater without limit!

THE PRINCIPLE OF EQUIVALENCE

Let us get back to the centrifuge experiment, and ask ourselves a simple question. Suppose the centrifuge is somewhere out in space, far away from the earth and its gravitational field. Suppose that the centrifuge is spun around so that the force is exactly one gravity. (This is represented by an acceleration of about 10 meters per second per second.) If you were in the chamber under these circumstances, you would feel right at home. Now suppose that you are sealed in the chamber, and that there are no windows in it, and no instruments that allow you to determine your whereabouts. How would you know that you were spinning around and around, and not just sitting back on the earth in a stationary chamber? There would be no way for you to tell the difference. The force of acceleration is just like the force of gravity. Einstein noticed this

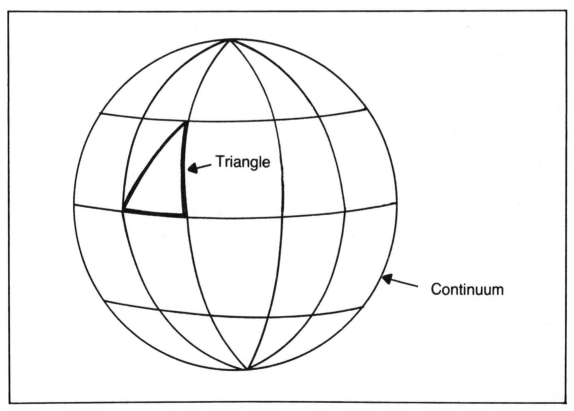

Fig. 4-6. The curvature of space on a large scale is believed to be positive, or spherical. The sum of the measures of the angles of a triangle is therefore larger than 180 degrees, provided the triangle is big enough.

resemblance between the two forces, and based his general theory of relativity on the proposition that they are actually the same! This is called the principle of equivalence. Whatever happens in the centrifuge chamber, accelerated at one gravity, will happen in a chamber situated on the surface of the earth. Conversely, whatever happens in a chamber on the earth will also happen in the centrifuge.

We have seen that extreme acceleration can cause the path of a light beam to be bent. According to the principle of equivalence, a sufficiently intense gravitational field should cause this same thing to happen. The intensity of a gravitational field must be tremendous in order to produce a distortion as great as that depicted in Fig. 4-4B; you would be crushed to death in such a situation, and the atoms of your body would degenerate into neutrons! But even in the earth's gravitational fields, some spatial distortion exists, if we are to believe the principle of equivalence. Scientists have detected this distortion in space near the sun. Two experiments have verified that the principle of equivalence is an accurate representation of reality.

Sir Arthur Eddington, an English astronomer, sought to observe the effects of space curvature in May, 1919, during an eclipse of the sun. Stars were photographed in the immediate vicinity of the sun during the eclipse, and these photographs were compared to photos taken previously, when the sun was not in the midst of the stars. The positions of the stars were not exactly the same in the two sets of photographs, and this discrepancy could be attributed only to the bending of the light rays from the stars. Einstein had predicted that this would happen, and the distortion was found to have just about the same extent as predicted. The light arriving from the distant stars must always traverse the shortest possible path through space; near the sun, that path is not quite straight!

The other experiment, showing the distortion of distances near the sun, was conducted by the radio astronomer I. Shapiro. Using radar, the distance to the planet Mercury was measured at intervals as the planet passed on the opposite side of the sun. The shape of the orbit of Mercury is accurately known, but the radar measurements showed an increase in the orbital radius as the planet passed behind the sun. Although this change amounts to only about 40 miles, it can be attributed only to space curvature. Radar beams must travel geodetic paths through space; near the sun, such paths are distorted.

How intense can gravity become? There is no limit. The more intense the gravitation, the greater the curvature of space. If gravitation becomes powerful enough, space can be literally torn apart! Then, all the familiar laws of physics are meaningless. Such is a black hole.

GRAVITATIONAL DEPRESSIONS

The distortion in space near the sun, observed by Eddington and Shapiro and predicted by Einstein, is similar to the funnel-shaped curvature illustrated in Fig. 4-5, except that it takes place in three-space rather than in two-space. Perhaps there are four-dimensional creatures, and to them, our universe seems to have no thickness. Those creatures would have no difficulty envisioning the curvature of space in the presence of a gravitational field. They might call the distortion of space near the sun a "gravitational depression."

Gravitational despressions exist around all sources of gravity. All matter produces gravitational effects. There is a gravitational depression around the earth, and a gravitational depression around the moon. Space is curved because of the gravitational field produced by your own body. In fact, all object possess gravitational depressions, right down to the tiniest elementary particle. And, as if that is not strange enough, the effects of all gravitational depressions are unbounded. The gravitational field of the sun becomes less intense at greater and greater distances, but it never totally disappears. The same is true of the gravitational depression surrounding a proton, neutron, or electron! Each and every particle in our universe contributes something to the gravitational environment of our cosmos.

For a given amount of matter, the degree of spatial distortion at a given distance does not depend on the physical shape or size of the object. A baseball might have the same mass as a piece of cheese; but the spatial curvature caused by either object is the same a mile away. But when a mass is very dense, the curvature in its immediate vicinity can become large.

If the sun were suddenly to shrink, we would notice no gravitational change here on the earth. But at the surface of the sun, the gravitational effects would increase. The spatial curvature would become greater and greater at the surface, as the sun got smaller. Eventually, the gravitational depression would be deep, indeed! Figure 4-7 shows, in a dimensionally-reduced sense, what would happen.

The German astronomer Karl Schwarzschild discovered something peculiar about the gravitational depression surrounding a shrinking, spherical object. Given a certain amount of mass, and compressing it without limit, will cause the gravitational depression to deepen until, at a certain defined radius, the depth becomes infinite! (In Fig. 4-7, this would be indicated by a "funnel" having an infinitely long "neck.") The greater the mass of an object, Schwarzschild found, the larger the radius at

which this happens. The formula for determining this radius is quite simple; if m is the mass of the object in kilograms, then the radius in meters at which the gravitational effects become infinite is given by multiplying the mass by 1.48×10^{-27}.

If the sun were to shrink until its diameter were just 3-½ miles, it would literally disappear! The earth would have to be compressed to the size of a grape for this to happen. A small object such as a baseball would have to be crushed to a size smaller than an atomic nucleus. These examples might at first seem ridiculous; how could such density ever be attained? Yet, it is believed that large stars, as they cool off and die, might collapse until they reach this state. The radius at which the gravitational depression becomes infinitely deep is called the Schwarzschild radius. When anything gets smaller than the Schwarzschild size, it must disappear. It must become a black hole.

THE HOMEOMORPHISM

Under ordinary circumstances, space can be bent and stretched to any extreme, but it can never be torn apart. Einstein used the concept of curved space to prove that the universe can be considered equivalent from all points of view. No matter whether you are moving, accelerating, rotating, or standing on the surface of a planet, the points in your universe can always be paired off, one-to-one, with the points in anyone else's universe. This pairing off of points, called a one-to-one correspondence, is called a homeomorphism. In the black hole, homeomorphism is destroyed.

Consider again the two-dimensional plane universe, and imagine the funnel-shaped gravitational depression. As long as the source of gravitation is larger than its Schwarzschild radius, the points in the universe surrounding the mass can be paired off one-to-one with the points in a flat plane (Fig. 4-8A). For every point in the curved universe, a mate can be found in the flat universe, and vice-versa. This pairing off, or homeomorphism, can be constructed in an infinite number of ways, since there are an infinite number of points! Figure 4-8A shows just one method of getting the correspondence.

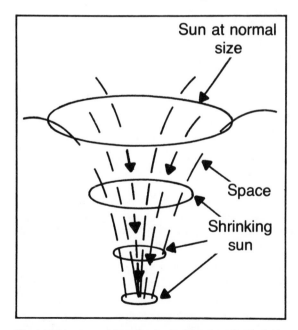

Fig. 4-7. As a star, such as the sun, collapses under the force of its own gravity, it is pulled into a space "warp," or gravitational depression.

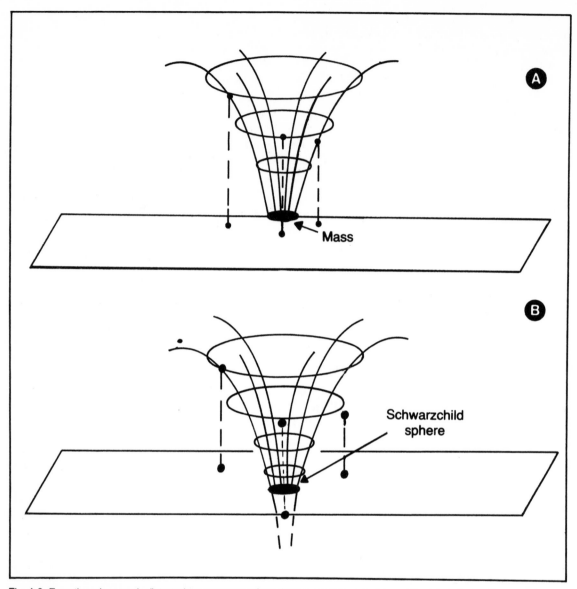

Fig. 4-8. Even though space is distorted near a source of gravitation, a homeomorphism still exists between the curved space and flat space. At A, this is shown by projecting the points vertically from the curved continuum to the flat. But when a black hole forms, the gravitational depression becomes infinitely deep, with a hole at its center, as shown at B. Inside the hole, there are no points, and a homeomorphism cannot exist.

The gravitational depression surrounding the mass may deepen, and the curved space will remain homeomorophic to a flat plane. The same thing happens in our three-dimensional universe in the vicinity of a shrinking, dying star. But when the star gets down to its Schwarzchild radius, the gravita-tional distortion becomes infinite, and the homeomorphism is lost (Fig. 4-8B). How does this affect reality? It means that the collapsing star has been severed from our universe! An infinite barrier has formed between it and us. Actually, this barrier can be crossed in one direction: We can jump into the hole. But nothing can ever get out of the hole. Once inside, we would be trapped forever!

At the entrance to a black hole, time as well as space is ripped asunder. Scientists have shown that, in agreement with the general theory of relativity, clocks run more slowly in a gravitational field. The more intense the field, the more slowly time will go. When the gravitational distortion of space becomes infinite, the same thing happens to time. This is perhaps the most bizarre aspect of the black hole: An infinite time span is compressed to a few seconds, minutes, or hours. A finite time span is compressed into a single instant. That instant can contain the seconds, minutes, and hours of another universe. A moment in a black-hole universe can contain the rest of our eternity!

THE TOMB OF A MASSIVE STAR

In order to become a black hole, a star must at first have a certain minimum amount of mass. Our sun does not seem to have sufficient mass to become a black hole. When the sun finally begins its final state of gravitational demise, it will become what is called a white dwarf, with a density so great that a teaspoonful of its matter would weigh tons on the earth! But the outward repulsion among the atoms of this dense star will keep gravity from pulling it down past a certain point. The white dwarf will gradually lose its residual energy, cool, and grow dark. The sun will then be about the size of our planet, but many times as heavy. Gravitation will be intense at the surface of this dead parent star. But not intense enough to cause the formation of a black hole.

Stars have, on the average, greater mass than the sun, and according to the calculations of the astrophysicists, a star only somewhat more massive than our sun will become a black hole upon its collapse. The exact amount of mass necessary to achieve gravitational oblivion is uncertain, because the workings of the interiors of dense stars are not yet fully understood. But eventually, the battle will rage between the force of gravitation and the repulsion among the atoms comprising the star. Gravitation gets stronger and stronger with increasing mass, but the force that keeps matter intact is unchanging. Thus, a massive enough star would have no way of escaping the fate of becoming a black hole.

At first, all of the electrons in all of the atoms of the collapsar would be driven into the nuclei, and this would result in what has been called a neutron star. A neutron actually consists of a proton combined with an electron; neutrons can sometimes be split into these two constituents, and the reverse process is also possible. A dense ball of matter, consisting entirely of neutrons, is often the end product of a massive star following a supernova explosion. This ball of neutrons is called a neutron star. Such an object may possess an extremely powerful magnetic field, and a rapid rate of rotation. This produces the electromagnetic radiation that so baffled astronomers in the 1960s that the objects were called LGMs (for little green men)! Now, of course, we call them pulsars.

But even the destruction of the protons and electrons may not be enough to ward off the final *coup de grace*, which gravity, in its unyielding manner, administers. The neutrons themselves may lose their identity, being crushed together like a conglomeration of warm candies. The neutron star then becomes one great neutron, measuring perhaps 10 to 12 miles in diameter, and having the mass of several suns. But then a curious thing happens. It can be explained from several different standpoints. The star just vanishes from the universe; it locks itself in a self-made time-space cell. Even light cannot escape the gravitational field of the star, and relativistically, the star kicks itself infinitely into the future. The star becomes a black hole.

All during the collapse of the neutron star to the Schwarzchild radius, the escape velocity has been increasing. Finally, it becomes greater than 186,282 miles per second, and the photons themselves, struggling to escape the overwhelming attractive force, fall back. Gravitation exerts an ever-increasing effect on the rate of time flow at the surface of the collapsing star. When the star falls within the Schwarzchild radius, time literally comes to a halt. For this reason, the sphere representing the Schwarzchild radius is called the event horizon. Beneath the event horizon, as far as anyone in the outside world is concerned, nothing ever happens.

The star now enters its own new universe. There is no chance of it ever returning to the old. It has ripped itself away. Anything that comes too near the black hole will be inexorably drawn into it, and since the star is so dense, the unfortunate wanderer will be crushed and assimilated into the star itself.

SITTING ON A COLLAPSAR

If we were to somehow land on a collapsing star and endure the intense gravitation present at the surface, what would we see as the star got smaller, ultimately retreating into the event horizon? We would have to be very astute observers to notice anything at all before the neutron material would crash down upon us from all directions. The end would come in a few seconds. But again, since we are only imagining this, we can, at will, slow down our perception of things by any factor we choose. Let us, in our minds, make milliseconds stretch into seconds!

At first, nothing peculiar would seem to happen, although the surface of the neutron star would appear very smooth. In the tremendous gravitation field of such a dense object, there could be no hills higher than a few inches. The scenario would be drab, indeed. But then the influence of gravity would begin to noticeably affect the paths of light rays coming to our eyes from the surrounding stars and landscape. The horizon of the neutron star, actually very nearby because of its small diameter, would seem to retreat. Then, oddly, the surface of the star would seem to be curving upward away from us, and we would get the feeling of sitting at the bottom of a huge parabolically shaped bowl. The stars in the sky would become distorted, too; we would see some of them rising upward from the horizon at all points of the compass. Only those stars originally at the zenith would remain fixed in our field of view.

As the gravitational distortion of space became more and more severe near the star, the horizon would move ever upward, and the visible sky would become ever smaller, although it would be filled with ever more stars. Finally, the sky and stars would recede to a single brilliant point of light at the exact zenith. A moment later, in a stunning flash, the sky would disappear completely. We would see only the surface of the neutron star. At this instant, photons of light traveling straight upward—the last to escape—would begin falling back. We might even see our own faces looking back down at us, no doubt with bewildered and fearful expressions!

At that time, we would have precious little life left. The hollow sphere of space, which would be our world, would grow tinier and tinier. Finally, with all of the violence of a whole star compressed nearly to a point, we would be crushed out of existence. The neutron star would have reached a singularity of space-time. It would become a single point, with no place to go, nor time to go anyplace.

THE VIEW FROM OUTSIDE

From the above example, it is evident that, even if our bodies could withstand the crushing force of gravity at the surface of a neutron star, we would die when the matter came down from all directions! Let us continue to suppose that we could live with the gravity; what would someone from the outside world see?

From the external point of view, there is no direct physical way to find out what happens to a collapsar once it falls inside its Schwarzchild radius. Time appears to go more and more slowly, and can approach, but never reach, that magic moment! A clock on the surface of a collapsing star would seem to tick more and more slowly until, ultimately, it would stop ticking altogether. That last second, just prior to the disappearance of the collapsar, would take forever to pass.

Actually, because of the relativistic effect on time, the light coming from a collapsing neutron star would grow so dim as to be invisible. The radiation would become so red-shifted that visible light would acquire kilometer wavelengths. Photons would be emitted so infrequently at the short wavelengths that our eyes would not see any light at all. But with sensitive enough apparatus, we would always be able to detect some energy coming from the black hole. If we had access to some kind of visible-light amplifier, which would allow us to see things a quadrillion quadrillion times less bright than we can normally see, then we might get a look at the travelers on the surface of the collapsing star. They would seem to be frozen stiff, in a state of suspended animation. The star would, likewise, seem unchanging. It would simply get dimmer and dimmer, but it would never disappear. Eventually, perhaps, only one photon would arrive at our vantage point every hour, every year, or every million years.

Somehow, the traveler into the black hole jumps off of the time line of our universe. Figure 4-9 illustrates an example of how time matches up between two reference frames, one on a collapsar and one outside it. The function is asymptotic—that is, the whole range of one variable is compressed into a defined part of the range of the other.

This brings us face to face with infinity. It gives the infinite a startlingly real aspect. You have probably heard the tale of the frog and the wall; the black hole gives this story a whole new significance! Suppose a frog sits facing a wall, and jumps exactly halfway to the wall. Imagine the frog repeating this process, always jumping half the distance of its previous leap. Will the frog ever reach the wall? We would say no. A viewer on a collapsing object would say yes! (We conveniently neglect the reality that the wall, the building of which it is part, and the earth on which the building rests, would all crumble in the meantime.)

The black hole creates a non-homeomorphism in time. As a collapsing object approaches its Schwarzschild radius, it is always possible to pair off the moments at its surface with the moments outside, until the object crosses the event horizon. That instant has no correspondent in our time universe. It never actually happens. In a sense, we can say that the object will pass through the event horizon in the infinite future (Fig. 4-9A). What does that mean?

THE JUMP TO INFINITY

The concept of infinity, or perhaps the lack of such a concept, has always intrigued and baffled the human mind. Some attempts have been made to define infinity, but the idea always seems beyond reach. But an event horizon gives infinity a starkly real aspect: If you jump into a black hole, infinity will be all around you. You will be thrown into the infinite future, and reduced to an infinitely small spatial volume.

When you were a small child, you might have boasted that you had counted to a thousand, two thousand, or five thousand. Such a task requires several hours. To reach a million, you would have to sit and chatter off numbers for weeks. To reach a billion, you would have to spend a lifetime. Larger numbers are beyond reach. But a billion is no closer to infinity than a thousand. Any attempt to strive toward infinity, by counting, means time wasted.

We have a system of writing numbers in terms of digital value. We call this technique the decimal system. We can count upwards in powers of 10, simply by adding zeroes to existing numbers. For example, we might count to a billion in a few seconds by saying "One, zero, zero, zero, zero, zero, zero, zero, zero, zero." That is a shortcut, and it skips most of the numerical values on the way from 1 to a billion, but it gets us there in less time. In a minute or so, we can reach extremely large numbers! But this method of counting is no better, in terms of approaching infinity, than counting by single integers.

Scientists can shorten numbers still further. We may write a 1 followed by a thousand zeroes as 10^{1000}, meaning 10 to the 1,000th power. We may then add another number 1,000 and enclose the resulting exponent in parentheses, obtaining 10 to the power of 1,000 to the 1,000th power: $10^{(1000^{1000})}$! That number, in conventional decimal form, would look like a 1 followed by 1000^{1000} zeroes. We could keep counting in this bizarre way, obtaining incomprehensibly large numbers. But this method of counting is no better, for the purpose of approaching infinity, than the previous two.

Division by Zero

You have probably heard of the idea that a fraction with zero in the denominator is "equal to

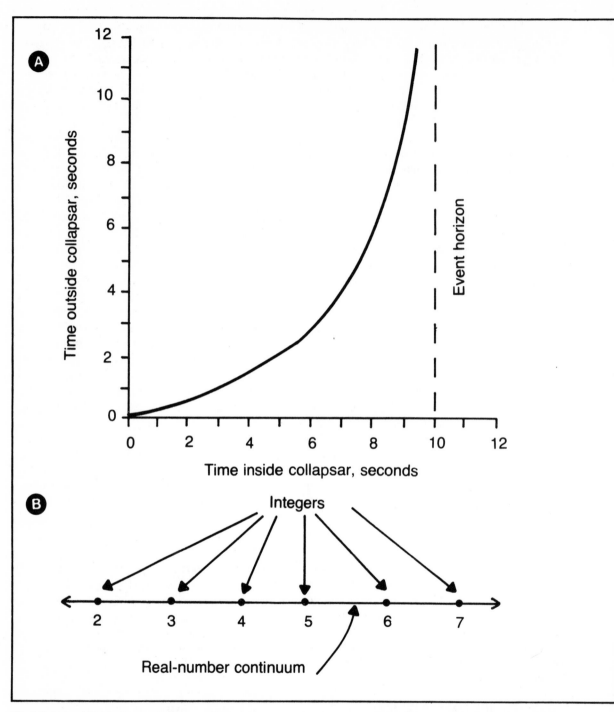

Fig. 4-9. At A, a black hole compresses an infinite time period into a finite time period—in this case 10 seconds. At B, integers appear as discrete points on a number line composed of real numbers. At C, a geometric transformation of an infinite length into a finite length. At D, a finite coordinate system, resulting from the geometric compression of an infinite volume.

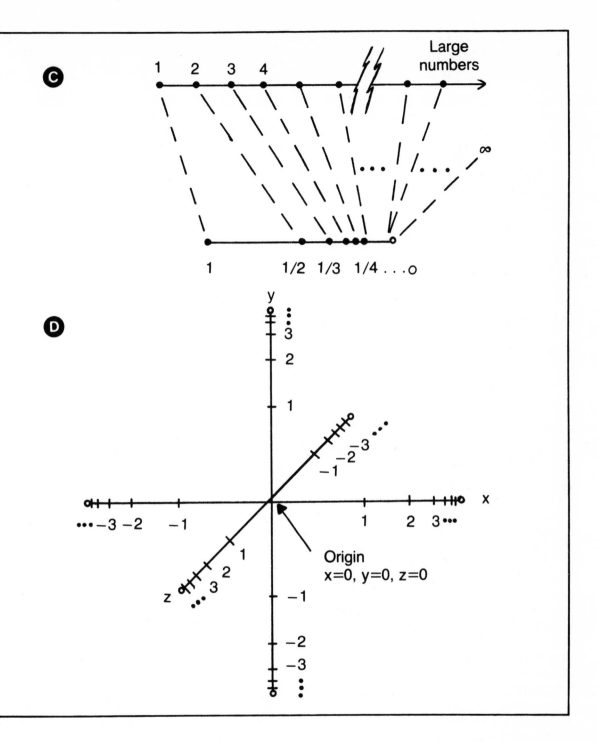

infinity." Mathematically, division by zero is simply not defined. That alone should make it interesting. Intuitively, it is appealing to say that $\frac{1}{0}$ is infinity, and we can give the newly "defined" number a symbol, the sideways 8. Therefore,

$$\frac{1}{0} = \infty.$$

If we dare to apply some of the simple rules of mathematics, we can investigate some of the properties of this "number." We can prove that

$$\frac{1}{0} = \frac{2}{0}.$$

In fact, we can prove that any positive number, divided by zero, gives infinity. That makes intuitive sense; infinity is the largest possible value. A whole mathematical theory structure can be built up around this concept. But eventually, if you undertake this little game for yourself, you will find that you are running in mathematical circles and getting nowhere. Mathematicians tried dividing by zero centuries ago. They got into trouble.

Aleph Nought

Set theoreticians are familiar with infinity to a degree perhaps not enjoyed by the followers of any other discipline. In set theory, the number of members in a set is called the cardinal number of the set. This provides an interesting opportunity to reach, intuitively, for infinity. Consider the set of grains of sand on all the beaches of the world. That is a large set! Its cardinal number is a large number! Consider the set of all the atoms in the known universe. That is a still larger set, with a still larger cardinal number. Consider the set of positive integers, or counting numbers: $\{1, 2, 3, \ldots\}$. That is an even larger set. Its cardinal number is bigger than any existing number—it is infinite. The set theoretician has given the cardinal number of the set of positive integers a name: aleph nought. To symbolize this value, which set theoreticians call a transfinite cardinal number, we use the capital Hebrew letter aleph with a subscript zero: $\aleph_0$.

At last, we can begin to understand a little bit about infinity. It is larger than any positive integer. You can never count to aleph nought. If you add a real number to aleph nought, you are left with aleph nought. The same holds for subtraction, multiplication, and division.

The odd thing about aleph nought is that, while it is bigger than any positive integer, aleph nought is not the biggest of the transfinite cardinals!

Aleph One

The integers are not the only numbers that exist. On a number line, the integers occupy individual, evenly spaced points, but the set of real numbers makes a continuum (Fig. 4-9B). A real number is, in a simplified sense, any number you can think of or make up, such as 1/2, 4.552, $\sqrt{10}$, or pi.

Is the set of real numbers the same size as the set of positive integers? The answer is no; there are more real numbers than positive integers.

How can this be? The set of positive integers is infinite; how can any set be larger? Mathematicians prove that the cardinal numbers of two sets are equal by finding a one-to-one correspondence between the set elements. If two sets are the same size, then it must be possible to pair off their members one-to-one. The converse of this is also true. No matter how we try to pair off the positive integers with the real numbers, however, we fail. There are just too many real numbers! If we assume that the sets are the same size, a contradiction invariably results.

The size of the set of real numbers must be defined by a new transfinite cardinal: aleph one (written $\aleph_1$). This cardinal number is infinite, but in a different sense than aleph nought is infinite. Intuitively, we reach a familiar impasse: Just when infinity seems to become somewhat comprehensible, we encounter its mystery again.

There might exist even larger transfinite cardinal numbers. We know there are at least two magnitudes of infinity; perhaps there are infinitely many. Which of these transfinite cardinals will confront us when we dive into an event horizon? The number of years, hours, minutes, or seconds between now and forever, or the number of instantaneous moments?

The Geometry of Infinity

The number line itself gives us another means of intuitively striving to grasp the concept of infinity. Every real number except 0 has a reciprocal, the value we obtain when we divide 1 by that number. For any real number x greater than 1, its reciprocal, $\frac{1}{x}$, is between 0 and 1. Also, for any real number y between 0 and 1, the reciprocal, $\frac{1}{y}$, is larger than 1. By choosing smaller and smaller values of y, closer and closer to 0, we can make x as large as we want. On the number line, every point larger than 1 has a mirror-image point—call it a cousin—between 0 and 1.

Suppose that we make use of this fact to compress the number line for values larger than 1 into a finite space. This puts infinity at a definite point on the line (Fig. 4-9C). You might recognize this little trick as a geometric expression of the division-by-zero idea!

We can do the same thing with the negative real numbers, and get a finite representation of the values more negative than -1. Then we can insert a normal real-number line, covering the values to -1 to 1, in between the two compressed number lines. This gives us a representation of the entire set of real numbers in the space of just four units.

Imagine, now, that we substitute these number lines for the infinitely long lines in the Cartesian system of Fig. 4-3. We obtain a cube (Fig. 4-9D) with edges measuring four units each, faces of 16 square units each, and a total volume of 64 cubic units. This cube contains infinitely many points. In fact, it contains just as many points as the entire coordiante system of Fig. 4-3!

With the coordinate system of Fig. 4-9D, we can map an infinite universe, in its entirety, within a finite space. We might represent every galaxy, every star, and every planet within the finite space 64 cubic units. For an earth-based coordinate system, we would assign our planet the point $(0,0,0)$: the center of the cube.

Naturally, we ask what lies outside of the cube. The answer: Infinity!

An Infinite Time Line

The illustrations of Fig. 4-9 can help us to develop an understanding of the nature of infinity in terms of space. A black hole affects space in much the same way as the coordinate transformations shown at C and D. But the black hole also distorts time to an infinite extent. The frog will reach the wall.

A geometric transformation can be used to get an idea of how it is possible to travel all the way to the end of time—and jump off! You have seen time lines in history books and inflation-rate charts. You have probably made your own time lines for personal-planning reasons. Figure 4-10A shows a linear time line of this sort. The line can begin at the present moment, some moment in the future, or sometime in the past: say January 1, 1900. To this point, we assign the value 0. We can give the subsequent points real-number values, such as 1, 2, 3, $5\text{-}\frac{1}{2}$, $7\text{-}\frac{29}{41}$, and so on. The units can be seconds, minutes, hours, years, centuries, or anything we like. The only requirement for a linear representation of time is that all of the units be the same size and the same duration.

We call the time line in Fig. 4-10A a linear representation of time, but by whose standards does time always go at the same rate? You might show Fig. 4-10A to a small child, who would then respond, "Says who?" Isaac Newton said something of this sort, and Albert Einstein replied, "Says who?" and proceeded to prove Newton wrong. We might represent time not as a linear quantity, but as a nonlinear dimension as shown in Fig. 4-10B. This scale is based on the geometric concept of the frog and the wall. Each unit on this scale is just half the length of the unit before. The value 0 is at the left end of the line; 1 is in the exact middle. At the right-hand end of the line, there is no correspondent in our time universe.

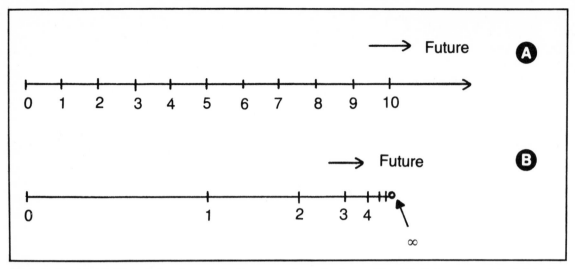

Fig. 4-10. We normally think of a time line as being linear, or uniform, like the rendition at A. But we can envision infinity if we compress, or "squash," the time line, as shown at B. The end point then represents eternity.

Mathematically, we can prove that the points on the line in Fig. 4-10A match up, one-to-one, with the points on the line in Fig. 4-10B, with the exception of the far right-hand point at B. This point is left without a mate at A.

Consider, now, that time moves along the line at B. A point representing "now" in our world would, then, travel at a continuously decreasing rate, like the frog jumping toward the wall. Our "now" would never get to the end of the line. But if we could jump into a black hole and watch a clock in the outside universe, that clock would seem to speed up so much that it would reach infinity. We would seem to travel along the line at B at a nearly constant rate, right to the end, and then we would fall off the end onto a new time line.

By venturing close to a black hole, it would be possible to travel far into the future. This kind of time travel can be accomplished because of the relativistic effect of gravitation on time. Whether we would ever want to make trips into the future is a debatable question; there would be no chance of a return ticket being issued. The same sort of future time travel is possible via high-speed rockets traveling at nearly the speed of light. But an infinite time trip, from our time line to another, seems vastly more terrifying than any journey into the future!

The black hole presents all sorts of fascinating possibilities to future travelers, should they ever be bold enough to leave our universe. We have seen that a collapsing star creates a gravitational field so intense that anyone who gets near the event horizon will be killed by the force. That would, at first, seem to preclude any chance of entering a black hole. But not all black holes need be as dense as a neutron star. According to the Schwarzchild equation, the event horizon for a conglomeration of matter is directly proportional to the mass. If you double the mass, you double the Schwarzchild radius. But if you double the radius of a sphere, the volume increases by a factor of 8. This means the density will be only one-quarter as great. You can show yourself, using simple algebra, that the density of an object at the Schwarzchild radius is inversely proportional to the square of the mass. What if 200 billion stars fell together? The resulting black hole would have a density 4×10^{22}, or 40 sextillion, times smaller than the density of a neutron star. An average-sized galaxy has about 200 billion stars, and there is good evidence that some galaxies are undergoing gravitational collapse. If a black hole has enough mass, its density will be far from formidable. It might even be comfortable!

DIVING IN

Imagine a hypothetical black hole so massive that you can fall within the event horizon without being crushed by the gravitational force. As you approach the event horizon, you do not notice any strange readings on your space-ship instruments. Looking out the windows, the black hole appears as a round empty ball, in the midst of a swarm of stars. Every now and then, a star seems to wink out as it approaches the entrances to the black hole; but you know, of course, that the star has only come to rest at the event horizon, like a grain of dust settling onto the earth.

Suppose that two space ships, yours and a companion vessel, make the journey into the black hole, one following the other. If you try to communicate with each other, you will find it difficult to keep track of the signals; they will fluctuate wildly in frequency. If your ship is leading, you will have to keep tuning your receiver to a higher and higher frequency; if you are following, you will have to go to longer and longer wavelengths. Not only will the frequency be altered, but the signal modulation will be accelerated or decelerated in such a way as to render a conversation utterly impossible. During that time while your two ships are on opposite sides of the event horizon, communications will be cut off in both directions. Your ships will, for that time, be infinitely far away from each other, in space as well as in time!

At the exact moment you cross the Schwarzchild boundary and thus make the transition from one universe to another, there is no "bang" or "bounce." There is no great revelation, except that the new universe suddenly becomes visible. You notice that there are individual stars in the new cosmos; in fact, the new universe looks, upon casual observation, quite similar to the old. But further observations must be made. The anatomy of the new universe must be ascertained. How large is it? Do any of the stars have planetary systems suitable for life? Are there galaxies of stars, as we knew them in the old cosmos? The previous universe must be forgotten. It has reached its end; it no longer exists.

It does not come as a surprise to discover that the new universe is a curved continuum, closed on itself in the configuration of a four-dimensional sphere. But the size of the new universe comes as a bit of a shock: The circumference is a mere 60 light years! That is microscopic, compared to the old universe, which was billions of light years around. You can see more stars at one glance than there really are in this cosmos! The new universe is so small that you can see around it several times. You might send a signal to yourself, and pick it up 60 years later! You might, if so predisposed, prepare to watch your original fall into the new universe! (You have 60 years to prepare your observing apparatus for the sight.)

The search begins, then, for a new planet on which to live. How many stars are there in the new universe? They are closely packed, like the stars in a globular cluster. Your scientists tell you there are 20 trillion stars! No doubt many of them have orbiting planets suitable for human beings.

The irreversible nature of a journey such as this, the fact that returning is out of the question, can never leave your mind. Your descendants died an infinitely long time ago; perhaps a more palatable way to say it is what they are living in another dimension of time, to which you can never return.

As you gaze back at the one-way door through which you have passed, you see, at intervals, new stars arriving. They seem to all come from a single point in space: a white hole.

Let us return to reality. You have made no such journey. But such trips may be possible. If large black holes, such as this hypothetical one, exist, we might enter and explore them!

DO BLACK HOLES EXIST?

The black hole "lives" only in the minds of the physicists and cosmologists, as far as we are presently concerned. No black hole has yet wandered into our solar system and fallen on the earth! (We hope that never does happen. Even the tiniest black hole will devour any object with which it collides.) But the search for black holes in space is being carried out by many scientists. Are black holes common? Are they perhaps as common as ordinary stars or planets? Or are black holes extremely unusual things? It seems that they must be quite commonplace, since apparently all stars having more than two or three times the sun's mass must end up as collapsars. An accurate idea of the distribution of black holes in the universe, or at least in our own galaxy, is essential if we are to draw conclusions about the ultimate fate of our universe.

Certain galaxies, in particular the radio galaxies, seem to produce far more energy than can be readily explained. It is thought that black holes might exist at the centers of such galaxies, where the stars are so closely packed that gravitational collapse is likely. Great amounts of energy would be liberated as matter falls into the abyss. Stars might be torn to pieces, and their hot interiors might be exposed, creating bursts of high-energy radiation. The immense red shift, caused by the gravitational field in the vicinity of the event horizon, could account for the radiation arriving here on the earth in the radio part of the spectrum.

Some astrophysicsts have gone so far as to speculate that there is a black hole at the center of almost every galaxy, including our Milky Way! This theory is supported by the observation of gravitational waves, apparently coming from the center of our galaxy. The huge black hole would pull in one star after another. With every new star, the Schwarzchild radius would increase slightly, and the curvature of space would be altered abruptly. This would create a shock wave in the space-time continuum, which we would notice as a gravitational wave. The effect is very similar to the ripples produced by a boat as a fish is caught and brought aboard.

It is unlikely that the actual collapse of a dying star will ever be directly observed. Such events may be frequent in our galaxy, but they last for a scant few seconds. The neutron star, already very faint, would seem to go dark in the twinkling of an eye! But perhaps someday, with computerized telescopes searching the heavens from a satellite or space ship, such an event might be recorded.

The quasars provide another possible instance of black holes in the distant reaches of space. All of the quasars are very far away, if we can believe the red-shift values we observe from them. Quasars liberate fantastic amounts of energy, but they are evidently small objects, no more than a light year across (and perhaps much smaller). About 1 trillion suns, packed together, would fall within an event horizon measuring a light year across. A typical galaxy has approximately this mass. Could it be that the quasars are black holes in the process of gobbling up the remains of entire galaxies? The immense power of gravitation seems the only factor that can explain the energy produced by the quasars. The search for gravitational waves arriving from quasars might provide clues as to whether they are black holes. Huge telescopes in space might resolve the quasars in enough detail so that we can look closely at them.

If the quasars are black holes, and they swallowed entire galaxies billions of years ago, we must presume that gigantic black holes are roaming among the galaxies in our vicinity, invisible because all of the stars are gone. If this is true, then the universe contains far more mass than we have estimated. As discussed in Chapter 1, that could radically alter our picture of the cosmos. It gives support to the oscillating-universe theory, and hope for the everlasting universe in which we would so like to believe.

Another phenomenon, the X-ray star, may prove to be associated with black holes. There are several different theories concerning the nature of the X-ray stars, which seem to produce more energy than is believed possible with ordinary nuclear reactions. One idea is that some X-ray stars are close binary systems in which one member has undergone gravitational collapse. Large amounts of energy would be liberated by matter torn from the visible companion and sucked toward the event horizon. Other binary systems have been observed, in which one star is not visible and yet has large mass. Such invisible stars could be ordinary black dwarfs or neutron stars; but they might be black holes.

Perhaps the search for the black hole has already reached at least a partial conclusion, however, in the theory that our entire universe is within its Schwarzchild radius. Some cosmologists have produced calculations showing that the density of our universe, in relation to its known radius, is of the correct magnitude for it to be a black hole. The question then arises: Shouldn't the universe be contracting, then, and not expanding, as it appears to be? The phenomenon of the black hole, and the things that take place inside it, are largely a mystery. Perhaps there is a limit to how small a black hole can become; there might be a balancing point. This balancing size would grow and grow, if a black hole were pulling in matter from the outside. Our universe could be such a monster, swallowing the matter from an unknown cosmic continuum from the infinite past.

USING BLACK HOLES

Any piece of material, regardless of how small it is, has a gravitational radius. Theoretically, it is possible to compress any object to such a density that it becomes a black hole. In practice, this may prove difficult or impossible. A small black hole could, however, to prove to be an immense, and efficient, source of energy. Can we take seriously the idea of manufacturing a black hole? Or might it be possible to utilize the energy of black holes already in existence? It is a far-fetched idea in today's world, but our technology has progressed so rapidly in the past half-century that, extrapolating into the future, such a feat might someday be well within the limits of our capability.

If black holes are ever harnessed and used as sources of energy, they will probably be allowed only in outer space. The small black hole might be contained by using strong electric or magnetic fields, thus suspending it in the center of an evacuated chamber, away from material objects. If the electrical system keeping the black hole in suspension happens to fail, then, only the space ship harboring the black hole will be destroyed! A black hole, even a tiny one, would be a perilous thing to have on the earth; if it got out of control for just a few moments, the whole planet would be annihilated! While this idea is far-fetched, it would (at least) make a great science-fiction subject, since it is a theoretically possible scenario.

When mankind develops space vessels capable of traveling at speeds approaching that of light,

perhaps journeys will be attempted in the far reaches of space, outside the solar system. Will black holes present a danger to space travelers? Probably not. The probability of colliding with even a visible star is practically zero. Space contains mostly that—space—and very little star, planet, meteor, asteroid, or other matter, except diffuse individual atoms and dust grains. Even if there are as many black holes as stars out there, the chances of hitting a black hole are smaller than those of hitting a star; collapsed stars are only a few miles in diameter. It is possible that black holes, since they cannot be seen, might cause unexpected course changes for a space vessel. Although invisible, their gravitational effects are as great as those of ordinary stars.

Gravitational fields can, in certain situations, be effectively used to increase the speed of a space ship. A black hole would be especially useful and effective for this purpose. If space travelers knew exactly where the black holes were in their vicinity, it is perhaps imaginable that the travelers would make a near pass on purpose. In this way, the tremendous gravitational energy of the black hole would be put to constructive use. Figure 4-11 shows basically how this is done; with careful navigation, a space ship could approach the event horizon, make one close orbit, and leave in the same direction from which it came, but with greater speed. This kind of maneuver would have to be done carefully, of course, because a small error could result in the ship being thrown far off course—or worse, pulled inside the event horizon. Certainly the accuracy of the ship's instruments would have to be refined to a degree much superior to the precision of what we have today.

The fabrication of a small black hole would require extreme expenditure of energy, in order to force a given amount of matter within its event horizon. But let us assume, for a moment, that we could make a small black hole, with a mass large enough to produce some gravitational effects. Such a black hole could conceivably be used to obtain propulsion for a space vessel already traveling at high speed. Space is not completely empty; there are some atoms out there, mostly in the form of

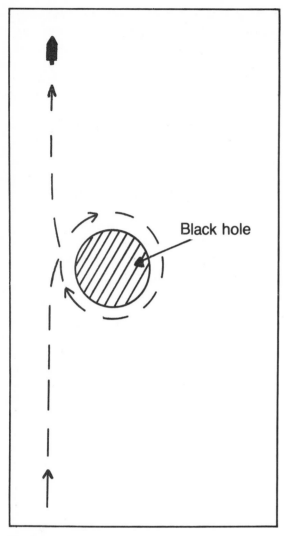

Fig. 4-11. A space ship might gain speed by closely orbiting a black hole. But the navigation would have to be perfect, or the ship would career off in an undesired direction. It might even be pulled inside the event horizon!

hydrogen, but also some heavier elements. A large collecting device could be used to scoop up the diffuse matter of interstellar space, funnel it down to a central point, and guide it into the voracious maw of the black hole! The ship would then gain speed because of the "suction" effect; matter would be drawn into the ship, but would never come out! The black hole would grow very slowly, and eventually it would become too large to handle. It could

then be discarded, and a new one put in its place. Again, by the technological standards of today, this sounds like a subject for science fiction, but let us not underestimate ourselves.

It is sad commentary for the human race that, if a black hole is ever actually manufactured, it will probably first be put to destructive use. This was the case with nuclear energy, and the unharnessed power of the black hole would make the largest hydrogen bomb look like a firecracker. To destroy the entire earth, it would only be necessary for a suicidal little madman to drop a small black hole. The singularity would fall back and forth through the earth as easily as a lead shot falls through water; earthquakes would first result as the tiny marauder cut its way through the crust and mantle of the planet. After a length of time that would depend only on the size of the initial black hole, the whole earth would be pulled inside an event horizon with a radius of about a half inch.

UNIVERSES WITHOUT END

Our own universe may be a black hole. This possibility has been admitted by some cosmologists. If this is true, then is it possible that there are other black-hole universes? It could be. If a large enough amount of matter gets packed inside its own gravitational event horizon, it will form its own dimension of time, and cut itself off from the rest of the cosmos, and it will have a density not hostile to our form of life. You have already made an imaginary journey from our universe into a smaller, black-hole universe.

If it is possible to have black holes in our universe, which is itself a black hole, then it ought to be possible to have an endless succession of black holes. There might be black holes within black holes within . . . Further, our universe could be a black hole with a larger black hole, inside of another . . . ad infinitum. That certainly is an interesting daydream. But the truth is, no doubt, still more bizarre.

It is only necessry to have one black hole in the universe, of sufficient size so that it has a habitable density, and an infinite succession of time dimensions can exist. For, if we dive into the event

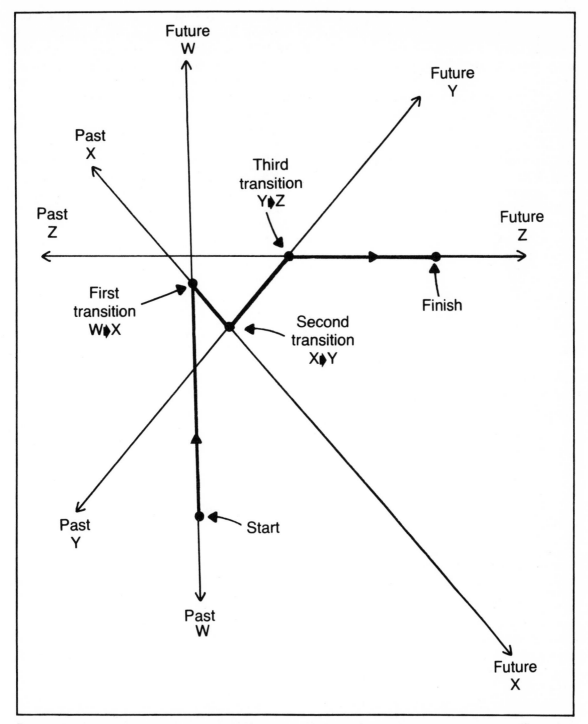

Fig. 4-12. A traveler through time dimensions might dive into one black hole after another, each time landing on a different time line. Here, three such leaps result in a journey through four time dimensions. The process could go on forever.

horizon, we are removed to the infinite future; we could dive into the hole again, the other way, and be thrown into a third dimension of time. But we would not find ourselves back home, in the old universe, if we reversed our direction. Another entirely new cosmos would await us, and we would be taking the risk that it might not be as benign as the previous one. In theory, we might dive back and forth in the black hole, through one event horizon after another, and encounter an unlimited potpourri of universes (Fig. 4-12).

Perhaps there is something about black holes that changes the seeming absoluteness of the event horizon. It could be that there is a missing factor in the Schwarzchild equations, that does not become noticeable until the strength of a gravitational field becomes so great that the black hole is formed. This was the situation with Newton's theories, and it could be also with Einstein's general relativity. Inside the black hole awaits the unknown, and only by actually plunging into a singularity will we ever find out—for certain—what it contains. A scientist, donning a space suit and climbing into a vessel bound for the ultimate threshold of an event horizon, might be regarded as a madman. But what is madness? A century or two ago, the very mention of a black hole, containing something beyond all that is, would have been regarded as insane. In a surreal universe, however, we must first realize that anything can happen. Only with such an imagination can we hope to keep uncovering the mysteries of the cosmos.

Chapter 5

The Frontiers of Cosmic Reality

NOT ALL OF THE PUZZLES OF THE UNIVERSE involve physical objects. Certainly, the physicist has much to think about in unraveling the mysteries of quasars, pulsars, and the nature of matter in general. What, in fact, is matter? Even the simplest object apparently is a collection of particles as vast, on its own scale, as any galaxy. And at the opposite end of the size scale, it is possible that the stars and galaxies are the elementary constituents of some other, unimaginably larger, universe. But on at least one planet in the cosmos, there exists a phenomenon with which the biologist can relate: life!

For several centuries, man has known that the planets are other worlds, comparable in size to our own planet. This fact has resulted in innumerable science-fiction stories about life elsewhere in the universe. Are we an accident? Or are there other intelligent civilizations? How might we find out and communicate with them?

In this chapter, we will investigate the possibility that there is intelligent life somewhere else in the universe. If our race should ever contact such a civilization, it would be the greatest moment in the history of mankind. We do not want to think that we are alone in the universe.

We shall also look at some of the most bizarre physical theories concerning the nature of matter, which, in our cosmos, is the link between minds, the mode by which we all communicate. Not all of the mysteries of the universe are large and far away; some are too small and nearby to be seen.

THE SOLAR SYSTEM

Compared with the remoteness of the stars and galaxies, our solar system is "our back yard." It is well within the realm of possibility that we shall someday travel throughout this planetary system. Even the most distant vlanet known, Pluto, is less than 40 times as far from the sun as we. With space ships traveling at the realistic speed of just 10 percent of the speed of light, we might reach Pluto within a few days. Thus, naturally, we should not be so far-sighted that we overlook our solar system as a possible abode for extraterrestrial life.

Imagine that we travel in a space vehicle, capable to reaching a speed of less than 1 percent that of light, and take a brief tour of the planets. What will we find? Certainly, some of the sights will be stunning. The variety of landscapes in the solar system is far greater than anything we will ever see on the earth. The deserts of Venus make Death Valley look like a lush paradise; the green clouds of Uranus present color combinations we never see on this planet. Let us begin from an orbit around the sun, and briefly visit each planet. Might any of these places be suitable for our survival? What about the evolution of life, even of the simplest form?

Mercury

We slowly widen our orbit around the sun until we are 36 million miles above its fiery surface. Our captain navigates us to within sight of the planet Mercury, and we enter into orbit around the reddish planet.

Mercury is like our own moon, except much hotter. We will probaby not find that life has evolved here; the temperature at the surface reaches over 800 degrees Fahrenheit, and there is little or no atmosphere. The gravity on Mercury is not intolerable; it is about 38 percent as strong as that on the earth. The day on Mercury is long—almost as long as its year. The planet takes 88 of our days to orbit the sun, and its day is equivalent in length to 58½ of our days.

As we land on the surface of Mercury, we must stay near the terminator, where the sun is on the horizon. This is best done near either pole. The nighttime side of Mercury sees temperatures plunge far below zero, but near the poles, the fluctuations in temperature as far less extreme. It is there that we would most expect to find life.

The vista reminds us of our moon. The landscape is barren, and devoid of any signs of life. Mercury does not seem to be the sort of place where life would evolve. The next expedition here will have more time to stay, and a thorough search can be carried out. But it is not likely they will find anything more than the tiniest microbes.

As we return to our parent vessel in the little landing shuttle, the resemblance between Mercury and the moon is indeed striking. The surface is pockmarked with craters. There has been little erosion here, indicating that Mercury probably never has had much of an atmosphere (Fig. 5-1). We pull out of orbit and head for the next planet, Venus—named after the Roman goddess of love and beauty. How beautiful will we find the planet?

Venus

Early theories concerning the planet Venus ranged from the idea that the planet is a swampy jungle, teeming with life, to a windswept desert or a soda-water sea. The truth finally became known, with certainty, when Russian probes landed on the planet and made undeniable measurements. Venus is covered with clouds, and through a visual telescope, the planet appears almost as featureless as a cue ball.

The clouds of Venus are nothing like the clouds we know on our planet. Our clouds are made of water vapor, but the clouds of Venus (Fig. 5-2) are composed of sulfur dioxide, carbon dioxide, and other gases that are deadly to our form of life.

We know enough about the surface of Venus that we are not foolish enough to attempt a landing. The temperature is as hot as that on Mercury, but that is not the primary danger. Venus is about the same size as the earth, and it does hold down an atmosphere. But that atmosphere is so dense that our little shuttle would be crushed under the pressure. We would need a submarine to reach the surface without being killed.

Nevertheless, we can get down below the clouds and take a look at the surface of Venus. As we descend through the clouds, the light around us becomes yellowish, then orange, and finally a dull red. Then we find ourselves under the cloud ceiling, overlooking a world bathed in eerie red light. The terrain is irregular; there are hills and mountains. There is no evidence of water, although we might see a few pools of molten lava. If life has evolved on this hellish planet, it must be of a variety much different from ours. A short look convinces us that we might as well not stay long. The outside temperature of our shuttle is reaching the critical

Fig. 5-1. Mercury appears much like the moon, but it is hotter—temperatures on Mercury exceed 800 degrees Fahrenheit, with the sun just 36 million miles away (courtesy of NASA).

Fig. 5-2. Venus appears as a featureless ball in visible light, but in ultraviolet light we can see details of the cloud structure. Venus is nothing like our planet (courtesy of NASA).

point. We must return to our vessel.

Neither Mercury nor Venus offer very acceptable environments for us, should we ever desire to live elsewhere in our solar system. We will bypass the earth on our way to Mars. The journey will take a few days, so we can settle down and get comfortable.

Earth

As we accelerate outward from orbit around Venus, our planet is already visible. The earth is nearly at full phase as we gaze at it, and it is the brightest object in the sky except for the sun. Although we will not stop at the earth on our way to Mars, we can make some observations about our home planet from this distance. We have time to think about our home.

The earth is just the right distance from the sun for the evolution of our kind of life. The earth's orbit fluctuates only about 2 percent; in January, the planet is 92 million miles from the sun, and in July, 94 million miles. The seasons are caused by the tilt of the axis, which is 23.5 degrees from perpendicular to the plane of the ecliptic. If the orbit of our planet were much more elliptical, or if the axis were much more tilted, we might find it an uninhabitable place.

The earth is just a little bigger than Venus, and the gravity is about the same. If our planet were much bigger, it might have turned out much like Venus, with a dense, noxious atmosphere. If our planet were much smaller, it might have ended up like its lone moon. As we come nearer the earth, we realize how lucky we are—what a perfect environment our home has.

Some parts of the earth are hostile to human life. If we were to come to the earth as visitors from another planet, we might land in a desert or at the south pole, and conclude that the earth is uninhabited. Signs of life are not readily apparent, even from a nearby orbit. Strange lights can be seen on the night side through powerful telescopes, but they might be thunderstorms. Would we recognize them as cities?

A sobering thought comes to our minds as we hurtle toward the orbit of our planet: We might overlook life, even sophisticated life, on another planet! This could happen to aliens visiting the earth, and it could happen to us when we visit Mars, Jupiter, or any other planet. The search for life must be thorough.

Earth's Moon

The earth has one lone moon. Relative to the size of our planet, the moon is large enough so that we may call the system a double planet. We are fairly certain that there is no alien life on the moon—although we now have bases there, and consider the moon inhabited by our species.

Our flight will take us part the far side of the moon, never visible from the earth. This half of the moon is still largely unexplored, since communications with earth-based stations is unreliable. Through a telescope, the far side of the moon appears heavily cratered. The temperature at midday often reaches more than 200 degrees Fahrenheit. At night, the temperature drops to far below zero. This means little in familiar terms, however—the moon is almost devoid of an atmosphere.

We approach the 93-million-mile point.

Mars

As we get farther from the sun, its light gets dimmer, and reaches a level to which we are well accustomed. Our captain tells us that we are passing the orbit of our home planet, and as we look out of the ship, we see a bright double object hanging among the unblinking stars. The brighter of the two is the earth. The less bright is the moon.

No planet, aside from the earth, has attracted the attention of the life seekers more than Mars, named after the ancient god of war. It seems that the word "Martian" is just as familiar in our minds as "Earthling"! This is not entirely without reason; scientists believe that Mars may support primitive life forms, and may at one time have been suitable for habitation by beings like ourselves.

As we enter into orbit around Mars, we can see a few craters (Figs. 5-3 and 5-4). The atmosphere of Mars is extremely thin, and meteors often strike the surface. Mars has about 38 percent as much gravitational pull as the earth. Mars has about

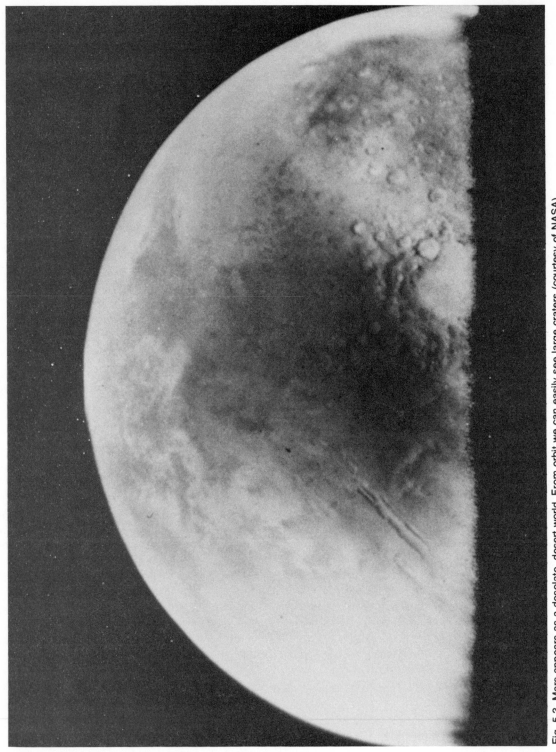

Fig. 5-3. Mars appears as a desolate, desert world. From orbit we can easily see large craters (courtesy of NASA).

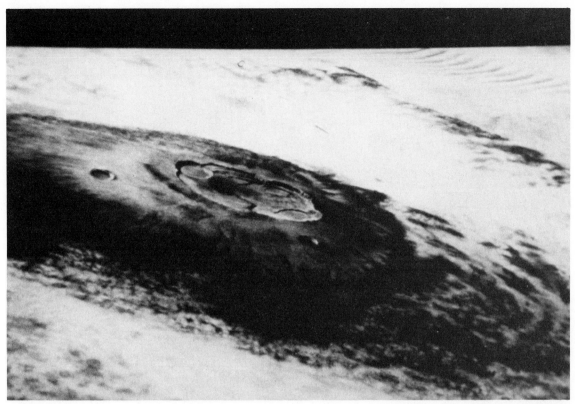

Fig. 5-4. Mount Olympus, the great Martian volcano, is 375 miles across and 15 miles high; clouds surround it, suggesting meteorological activity (courtesy of NASA).

the same degree of axial tilt as our planet, and the day on Mars is just over 24 hours in length. It's almost too familiar!

As we land on the surface of Mars, we are in the "tropics" at a latitude of about 20 degrees north. It is the height of summer, and the late-morning sun shines down with a surprisingly bright light (Fig. 5-5). Mars is 1.5 times as far from the sun as Earth, and it receives less than half as much sunlight. But our eyes perceive sunlight in a nonlinear manner. The difference between Mars and Earth is hardly noticeable.

The temperature on the surface of Mars can reach about 80 degrees Fahrenheit during a summer afternoon, but at night it gets terribly cold. Even at the warmest time of year, the pre-dawn hours bring a decided chill to the air—more than 100 degrees below zero. In this rugged climate, we might expect to find lichens or bacteria.

No life has yet been found on Mars, and we see no evidence of it from our vantage point. But the search will continue. Much of the surface of Mars remains unexplored. Mars may have had, at one time, a much thicker atmosphere than it now holds; the reddish color of the planet, in fact, consists of metal oxides, suggesting that Mars once had an atmosphere rich in oxygen. Perhaps someday we will find a way to unlock the oxygen from the rocks, and put it back in the air. Plans for such a project have been suggested.

Mars is the most distant of the inner or terrestrial planets. Until now, the planets we have seen are rather small. The next planet, Jupiter, is the largest of the known satellites of our sun. The trip to Jupiter will require over six months. But the journey will not be without some excitement. For, along the way, we must pass through the asteroid belt.

Fig. 5-5. The surface of Mars indeed suggests a desert world. But life may still exist, as it does in similar-looking parts of the southwest U.S. (courtesy of NASA).

The Asteroids

The asteroids, sometimes called planetoids, are literally millions of pieces of rock, in orbit around the sun between Mars and Jupiter. Scientists have several theories concerning the reason for the presence of asteroids, rather than a planet, in this region. Probably, the gravitational field of Jupiter prevented the formation of any planet here. Or, perhaps, a planet once did exist, but tidal forces from Jupiter tore it to pieces.

If all the known asteroids were put together, they would form a small planet, about the size of our own moon, not large enough to hold an atmosphere. This, at least, provides some reassurance against one chilling theory: that an inhabited, advanced planet was shattered by bombs manufactured by its own people. There is still some uncertainty about the total mass of all the asteroids. Perhaps our

estimate is grossly in error. We hope not!

The hazard of the asteroids for our ship, however, is of immediate concern. We pass through this region on a special course, plotted to minimize our velocity with respect to the majority of asteroids. Just one of these pieces of flying rock, hitting our ship at a speed of 1 mile per second, would end our journey instantly. The radar must be constantly watched!

Jupiter

As we travel on, one object in the sky becomes brighter and brighter. Even from a great distance, Jupiter is spectacular (Fig. 5-6). Finally, we enter into a high orbit around this, the largest of the planets (Fig. 5-7). Jupiter is named after the greatest of the Roman gods, and as we look upon the planet, we know that the name is well chosen.

Fig. 5-6. Jupiter is spectacular, even from hundreds of thousands of miles away (courtesy of NASA).

Fig. 5-7. Jupiter's clouds loom from a lower altitude. Moons are visible in the foreground (courtesy of NASA).

Jupiter is 89,000 miles in diameter, or about 11 times the diameter of the earth. The gravitational pull on Jupiter is more than twice as strong as that on earth. Jupiter is 5.2 times as far from the sun as we, and it gets only 4 percent as much light and heat from our parent star. Despite this, the clouds in the atmosphere of Jupiter are bright, because they reflect so much light from the sun.

We will make no attempt to land on Jupiter. First, we do not know that it even has a surface of the type that would allow a landing. Second, the intense pressure, deep within the planet, would probably crush our shuttle. The clouds of Jupiter are deep, dense, and consist largely of methane and ammonia. While these gases are poisonous to us, they are favorable for the formation of amino acids, from which life as we know it arises.

Scientists believe there might be primitive life within the clouds of Jupiter. The farther down we go into the clouds, the warmer the temperature becomes, and at some point, we might find bacteria or perhaps even more complex organisms. If life has had the chance to evolve to any level of sophistication, we might see birds or floating fish-like creatures living in the atmosphere.

The atmosphere of Jupiter is complicated, and is divided into several different convection bands. Storms of incredible violence continually rage (Fig. 5-8). The turbulence occurs largely because Jupiter has such a rapid rotational rate. The day is less than 10 of our hours in length, and the resulting Coriolis effects are tremendous. Some of the storms probably contain lightning discharges. It is thought that this increases the chance of the formation of amino acids, and perhaps of elementary life forms.

We will land on Ganymede, Jupiter's largest moon, and the biggest planetary satellite in the solar system.

Ganymede

Ganymede is larger than the planet Mercury. As we land, however, we are somewhat disappointed: There is scarcely any atmosphere. There is evidence of water; Ganymede is believed to be composed of ice and rock. While it is unlikely that we will find life here, we can get a stunning view of Jupiter only 670,000 miles away. Ganymede is an excellent place for setting up complex apparatus to observe Jupiter in detail.

Volcanic eruptions occasionally take place on Ganymede; the planet has a warm interior. The abundance of water ice is perhaps a good sign for the survival of man here, once the sun gets too hot for life on the earth. At that time, Ganymede may present a hospitable—although not lush—abode for us.

Jupiter has many other moons; three of them are large enough to be seen from the earth through a small telescope. But none of these satellites are places on which we should expect to find life. As we leave Ganymede, and prepare to continue to Saturn, we feel a sense of remorse that we could not have stayed, and taken a journey into the Jovian atmosphere. That will have to be left up to future visitors here, with spacecraft designed to withstand the violence of stormy Jupiter.

Saturn

Shortly after accelerating out of the gravitational influence of Jupiter, Saturn appears as a bright star in the sky. As we approach this planet, we notice that it appears somewhat elongated. Of course, this is because of the bright ring system

Fig. 5-8. The Great Red Spot is thought to be a storm in the atmosphere of Jupiter. Resembling a gigantic hurricane, it is larger than the earth. Atmospheric currents can be clearly seen surrounding the Great Red Spot (courtesy of NASA).

Fig. 5-9. Saturn as seen through a telescope from the earth (courtesy of Mount Wilson and Las Campanas Observatories, Carnegie Institution of Washington).

that surrounds the planet; finally, we are close enough to be able to resolve the rings with unaided eyes (Figs. 5-9 through 5-11).

Saturn is almost as large as Jupiter. In many ways, Saturn is like a twin of the largest planet. The atmosphere of Saturn is evidently less violent than that of Jupiter, and the gravitational pull is about the same as that on the earth. We will descend at least part of the way into the atmosphere of Saturn, but we are prepared to turn back quickly if signs of trouble develop.

After stabilizing in an orbit well outside of Saturn's impressive ring system (Fig. 5-12), we begin our trip down. A high, thin layer of clouds comes first; the sky appears milky as we drop

below this upper cloud deck, and the sun acquires a ring that looks almost familiar. The panorama below is one of dark, twisting, billowing clouds. We decide that it would be too risky to enter them.

From one especially dark rift in the clouds, a lightning flash appears, and then another. Since the atmosphere of Saturn is very similiar, in terms of chemical composition, to that of Jupiter, we can expect a good chance of finding amino acids here. But Saturn is colder than Jupiter, and if life can evolve here, it would probably occur deeper in the thick atmosphere. Saturn probably has a surface of oozing, liquid metallic hydrogen, shrouded in eternal darkness because of the thick clouds.

Saturn itself is too dangerous for our craft—

observations indicate winds in excess of 300 miles per hour, comparable to a typical tornado on the earth—but high in the sky is a small speck of light that has proven to be one of the most interesting places in the solar system. As we gain altitude, we recall the name of this satellite: Titan.

Titan

The most interesting feature about Titan, one of the moons of Saturn, is that it has a thick atmosphere. In fact, the atmosphere of Titan is thicker than that of the earth, and the most abundant element is nitrogen, just as in the atmosphere of our home planet. Titan is almost completely covered by orange clouds. We enter into orbit around this small planet.

Some scientists have envisioned Titan as the most likely abode for life in the solar system, other than the earth! But closer observations have revealed that Titan is terribly cold. Temperatures are always more than 100 degrees below zero; typical readings are −250 to −300 degrees Fahrenheit. Besides nitrogen, the atmosphere of Titan contains methane, perhaps some hydrogen, and other substances similar to gasoline. As we descend toward the cloud layer, Saturn is spectacular at a distance of 750,000 miles! The sky appears slightly hazy overhead a reminder that Titan actually has an atmosphere.

The view from the surface of Titan is far less impressive. The sky is always a dusky brown-gray; rarely does the sun peek through the overcast. Volcanic activity is frequent, and these hot spots are the most likely areas in which amino acids might form. Titan, like Ganymede, may be ready and waiting for us when the sun begins to expand into a red giant. But at the moment, Titan is in a state of icy, suspended animation: The ingredients for life

Fig. 5-10. Saturn as seen by an American Voyager space probe (courtesy of NASA).

Fig. 5-11. Saturn as it would appear to an approaching spacecraft (courtesy of NASA).

are there, but the temperature is simply not warm enough.

We blast off from Titan, with its ruddy complexion, and prepare to leave the vicinity of Saturn. We realize that the chances of finding life on the three outer planets—Uranus, Neptune, and Pluto—are not good. They are too far from the sun, and too cold, for the evolution of life to be likely. But we must not get too narrow-minded; there may be biological bases for life that are much different from ours. Could it be that we might find creatures that set their thermostats at −325 degrees Fahrenheit, breathe poisonous gases, and swim in pools of liquid nitrogen?

Uranus

Aboard our space ship, there is no real day or night. We must create our own cycle, 24 hours in length, in our minds. The sunlight is constant. But, at the remote distance of a billion miles from the sun, the light is noticeably dimmer than the level to which we are accustomed. A sunbeam slanting through the window of our ship is less than 1 percent as brilliant as on the earth.

The distances among the outer planets are much greater than the separation among the inner planets, and this fact is tediously evident now. The journey from Saturn to Uranus is months long. But finally, a green disk appears in the sky.

Uranus is somewhat smaller than Jupiter or Saturn, but its atmosphere is similar. Faint, hazy bands appear on the disk. These are cloud bands. Weather systems exist on Uranus, but they are less violent than those on Jupiter or Saturn.

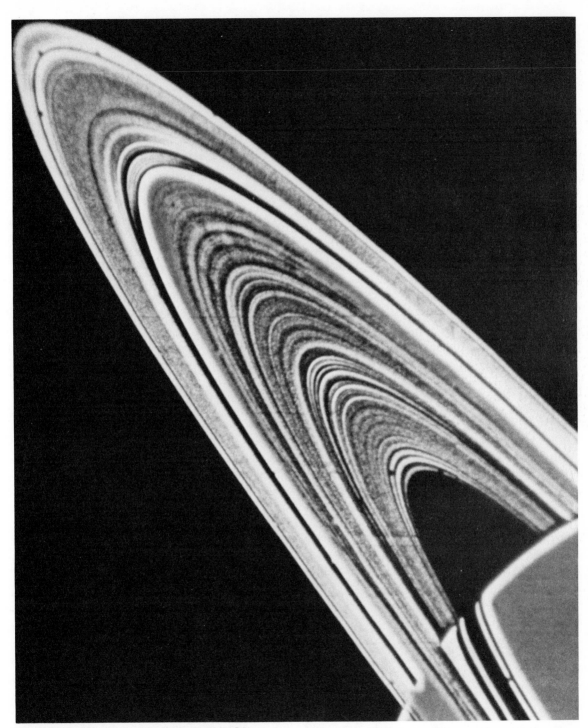

Fig. 5-12. Saturn and its rings (courtesy of NASA).

The most interesting thing about Uranus is the intensity of its seasons. The other planets have tilted axes, but not to the extent of the tilt of Uranus' axis. Our own planet has a tilt of 23.5 degrees with respect to the plane of the ecliptic, and that is quite enough to produce dramatic seasons! Uranus lies at an angle of almost 90 degrees. First one pole, and then the other, is exposed to the sun.

The temperature on Uranus is always about 300 degrees below zero. From the surface, the sun would appear dim or invisible because of the clouds. The chances of life here, in a form familiar to us, are small. Uranus may have a solid surface because of the extreme temperature, but it is more probably a sluggish sea of liquid hydrogen, resembling the environments of Jupiter and Saturn.

Neptune

Neptune is yet a billion miles farther from the sun than Uranus, and therefore it is even colder. Neptune looks very much like Uranus; it has somewhat bluer hue, but it is approximately the same size. Neptune is named after the god of the sea, because of its blue color. And in fact, the entire surface of Neptune is probably a vast ocean, but the gods would find it a dull ocean of liquid metallic hydrogen.

The atmosphere of Neptune is much calmer than that of Uranus, and the gravity at the surface is believed to be only a little more intense than that on the earth. There are occasional clearings in the clouds, and we decide to make a landing if possible.

The atmosphere of Neptune has abundant hydrogen and methane, like the atmospheres of Jupiter, Saturn, and Uranus. We lose altitude, and the sky becomes indigo and then blue. The scene is remindful of a view from an airliner on the earth. Below us, a flat, blue-gray ocean stretches from horizon to horizon, interspersed with billowy clouds.

The hydrogen ocean rolls gently, in huge rounded swells. There are storms on Neptune, but we have cautiously avoided them. But the temperature—350 degrees below zero—precludes any chances of finding life as we know it.

Pluto

Pluto is, for much of its orbit around the sun, actually closer to our parent star than Neptune. Pluto is a small planet, believed to be about the size of our moon. Pluto has a large companion, Charon. Neither of these planets is believed to have any atmosphere. There is evidence of frozen methane on the surface of Pluto.

From a vantage point on the surface of Pluto, the sun casts an eerie dim light on the bleak rocks. Pluto gets about 0.06 percent as much light as the earth does. The chances for life evolving in this environment are essentially nil; the planet is a continual super deep freeze.

This is the end of our imaginary journey through the solar system. The trip back to our home planet is a long one. If our descendants ever make a trip such as the one described here, they will no doubt feel elated when the blue-and-white disk of the earth finally appears.

The above "mind journey" is full of question marks, because no man has yet visited any of the other planets in our solar system save the moon. Unmanned probes have indicated that we might want to visit Mars, Jupiter, Saturn, and Titan; if we get the technology, we might also go to Uranus and Neptune. But no definite signs of life elsewhere in our solar system have yet been found. We may someday find bacteria, viruses, and perhaps small insect-like creatures in the solar system, but the evolution of intelligent life on any of our neighboring planets has probably not taken place. If we are looking for extraterrestrial humanoids, we will have to travel farther.

THREE METHODS

In the search for other intelligent life in the universe, we might employ three different techniques. These are:

(1) Try to visit them
(2) Try to communicate with them
(3) Wait for them to come to us

Each of these plans has some good and bad points. Interstellar and intergalactic travel is fraught with hazards and difficulties. Communication is a trial-and-error process. And just waiting . . . well, that is an easy way out, but it is undoubtedly also the slowest.

Interplanetary travel, within the limits of our solar system, will probably be a reality some day. It is not necessary to attain fantastic speeds to travel to any of the planets; nonrelativistic velocities would prove entirely adequate. But the situation is much different for any who attempt to reach the stars.

At the time of this writing, we have put a man on the moon. If the earth were the size of a marble, the moon would, on that scale, be another (smaller) marble a little more than a foot away. At present space-ship speeds, the trip requires over two days each way.

On the above scale, the sun would be about 400 feet away—roughly the distance from home plate to the center-field fence in a typical major-league ball stadium. The distance from the sun to Pluto would be about 3 miles. Yet, we do have the ability to span these distances!

But the nearest star, Proxima Centauri, would be incomparably farther away: 20,000 miles. Even if we are able to build a ship that can accelerate to half the speed of light, the round trip to this star system would take 18 years.

Our Milky Way galaxy is about 25,000 times as wide as the distance from our sun to Proxima Centauri! It becomes obvious that, if we are ever to attempt interstellar travel on a galactic scale, we will have to reach speeds at which relativistic time dilation takes place to a large degree. Let us momentarily dispense with the technological problems involved in getting a space vessel to travel at such speed. Let us even dismiss the problems of navigation at near-light velocity. Suppose a super-speed space vessel were awaiting us at this very moment, and within a couple of hours we could drive down to the spaceport, board a shuttle, blast off, and get aboard a starship bound for the other side of the galaxy. Would we do it?

With a space vessel capable of accelerating indefinitely at 1 gravity, or about 10 meters per second every second, we could reach relativistic speeds within a few weeks or months. Within a lifetime, we might travel to the other side of the galaxy and back, or even perhaps to another galaxy, such as the Great Nebula in Andromeda. But upon our return to the earth, we would find a planet that had aged hundreds of thousands, or even millions, of years! For those travelers leaving for the other side of the galaxy, or even for stars within a few tens of light years, the ugly reality must be faced: Everyone they know will be dead when they return. For many, this price will be too great.

Because of the difficulties of interstellar travel, some scientists have suggested that a better way to contact any possible extraterrestrial civilization, without the expense (both fiscal and psychological) of travel, is to attempt communication by electromagnetic means. This might include the use of radio transmitters and laser beams. It is believed that other civilizations may even now be attempting to contact us with radio transmitters. The main questions are: On what frequencies are they so doing, and on what frequencies ought we to transmit so they will be most likely to hear us?

The final alternative, that of waiting for aliens to come to us, is probably out of the question for a race as inquisitive as ours. There is some evidence that our planet may have already been visited by extraterrestrials. But conclusive proof is lacking. It is intriguing (and a little frightening) to think that someone out there might be watching and waiting for the right moment to land in our major capital cities.

Starship Design

If we intend to travel among the stars—at least the nearer ones—it will be necessary to build a space ship capable of reaching extreme speeds. It will also be essential that the ship support life for a long period of time, and provide an environment that is tolerable for the passengers. Many ideas have been discussed and proposed for starship design.

Whatever sort of propulsion system is used in the interstellar vessel, one thing is certain: It will have to be enormously powerful. Nuclear reactions are the most powerful source of energy that we can currently use. One design, invented by Freeman Dyson, Theodore Taylor, and others, and considered at one time by the United States, could employ hydrogen fusion bombs, exploded one after another against a large deflecting shield (Fig. 5-13). This design has been given the name of Project Orion. The fusion reaction has not yet been harnessed in an atomic reactor, but fusion is far more efficient than fission. Eventually, nuclear fusion will probably be achieved in a controlled environment.

Acting on the assumption that a fusion reactor will be perfected some day, the British Interplanetary Society has designed a starship they call Daedalus. Instead of using bombs, which provide bursts of thrust, the Daedalus ship employs a continuous fusion reaction, like a miniature sun, to give a smoother ride (Fig. 5-14). As with the Orion design, a large deflecting shield would be necessary to protect the passengers from the heat and harmful radiation.

Both of the above starships could reach speeds of about 10 percent of the speed of light. This is not fast enough to cause a large amount of relativistic time dilation. Such a speed is insufficient to reach any but the nearest stars within the span of a human lifetime. It is possible, however, that space travelers may continue their journey for generations. Some people might be born in space, live their entire lives in space, and ultimately die in space. There are obviously some far-reaching philosophical considerations, as well as physical hurdles, to be reckoned with in long-distance space travel!

A third method of achieving high speeds, actually approaching the speed of light, has been suggested by R. W. Bussard and others. A primary difficulty with the Orion and Daedalus vessels is that they must carry their fuel, which itself has great mass. Perhaps the diffuse interstellar gas and dust can be scooped up and used as fuel for a fusion reactor. With a big enough scoop, sufficient matter could be obtained, in theory, to operate the propulsion system (Fig. 5-15). This design has been called the Bussard Ramjet, since it works something like a ramjet. The faster the speed of the Bussard Ramjet, the better it will work. An acceleration of 1 gravity

could be maintained indefinitely. Within less than a year, the speed of the starship would be almost the speed of light. With such a space ship, the entire known universe is within reach. It would be possible for a traveler to go to, and come back from, any galaxy of his choice. Unfortunately, however, he could never hope to return to the same earth! While a few months or years might pass for a traveler aboard a Bussard Ramjet, the earth would age thousands, millions, or billions of years. It is possible that an intergalactic adventurer would return to find the sun a fading white dwarf, and the earth either vaporized from the red-giant phase or frozen at hundreds of degrees below zero.

Of course, there may be some form of propulsion other than nuclear power, such as matter-antimatter reactions, gravitational energy, or something altogether unknown at this time. It might be possible to avoid the relativistic effects that, according to our present knowledge, will haunt the thoughts of every aspiring deep-space traveler. If the extraterrestrials are watching as we ponder the means by which we will reach the stars, they may be marveling at how we have missed the obvious solution!

If we do travel to the stars, we will not be able to afford to search randomly for other civilizations. There will have to be some good evidence that we will find what we are looking for. How will we do this? One method is to search the heavens for evidence of radio signals from intelligent beings. Some scientists believe that we are being called at this moment by one, or even several, other intelligent societies.

Communications

The earliest wireless transmissions, conducted with spark-gap transmitters at low frequencies, probably did not escape the effects of the ionosphere of our planet. But shortwave radio signals can sometimes penetrate the upper atmosphere and continue into space. Signals at very high frequencies invariably pass through the ionosphere. By now, some of our earliest very-high-frequency broadcast energy has traveled past numerous stars in our part of the galaxy. Has any-

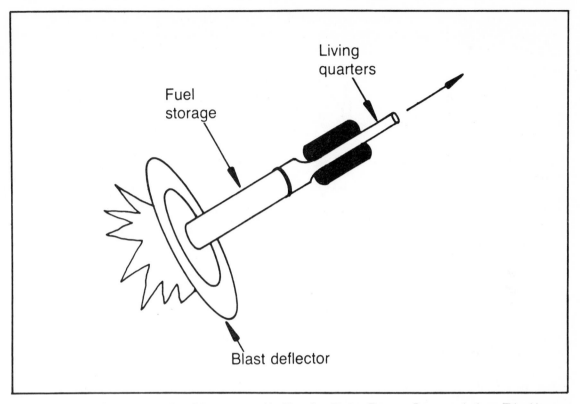

Fuel
storage

Living
quarters

Blast deflector

Fig. 5-13. The intestellar vessel called Orion, proposed by Theodore Taylor, Freeman Dyson, and others. This ship uses nuclear bombs for propulsion.

one heard them? Might they try to send something back to us? Could some of these signals have already fallen on our planet undetected by us?

The first serious attempt to find signals from another civilization was initiated by Frank Drake at Green Bank, West Virginia, in 1959. Drake and his colleagues called the endeavor Project Ozma, named after the imaginary land of Oz. The scientific establishment regarded Project Ozma with a combination of interest, amusement, and skepticism. However, Drake pointed out that, if enough stars were scanned with the sensitive radio telescopes at Green Bank and other radio observatories, results might well be obtained. A single confirmation of intelligent life on another planet would justify all of the expense and time invested—a hundred times over!

Some of the stars that Drake investigated were Tau Ceti, in the constellation Cetus, the Whale, and Epsilon Eridani, in the constellation Eridanus, the River. Various frequencies were checked, in the vicinity of the hydrogen resonant wavelength of 21 centimeters. The results were negative, but the allocated time was limited. Drake advocated that further experiments be conducted whenever a radio telescope could be made available.

Part of the problem in the search for signals from other civilizations lies in the fact that only a very tiny part of the sky can be scanned at any given time. This narrow field of reception is necessary because of the large amount of radio noise generated by celestial objects. The problem has a reverse aspect: Anyone attempting to transmit signals over interstellar distances must use a narrow signal beam. This is simply because an omnidirectional transmission would die out, and fade into the background of galactic noise, much more rapidly than a sharply beamed signal. The operators of the

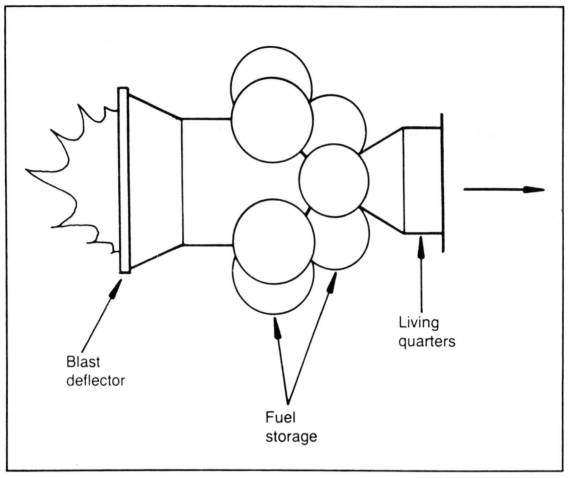

Fig. 5-14. The intestellar spacecraft called Daedalus uses a nuclear-fusion reactor for its propulsion. This design has been proposed by the British Interplanetary Society.

interstellar transmitter must direct the beam at certain particular stars. Which stars will they choose? Perhaps not our sun. The probability of our hearing the radio signals is therefore dictated by the square of the proportion of stars actually transmitting radio signals: It is as unlikely that they are transmitting at us, as that we will happen to point our receivers at them.

The antenna aiming problem is further compounded by the fact that the distant stars are not now where they appear to be. All of the stars are moving with respect to each other; the position of a star when its light leaves it, as compared to its position when a signal arrives from a distant trans-

mitter, might change considerably. This is illustrated in Fig. 5-16. The transmitted beam must be made wide enough to get rid of the possibility that the radio signals might miss their target! But this means that the transmitter itself must generate a tremendous amount of power, if the signal is to reach its destination.

Another problem with finding the signals from other civilizations concerns the frequency at which we should conduct the search. The 21-centimeter hydrogen resonant frequency, a natural marker in the electromagnetic spectrum, has often been suggested as the most likely wavelength at which intelligent beings would attempt to send signals to other star systems. However, this idea could be wrong; this frequency may be the one they consciously avoid because of the high noise level! Then, at which frequency should we listen? For radio signals to penetrate great distances in space, they must be concentrated within a very narrow range of frequencies, like a broadcast signal. It is not practical to transmit broad-spectrum electromagnetic energy, because the background noise will drown it out before it gets even 1 light year

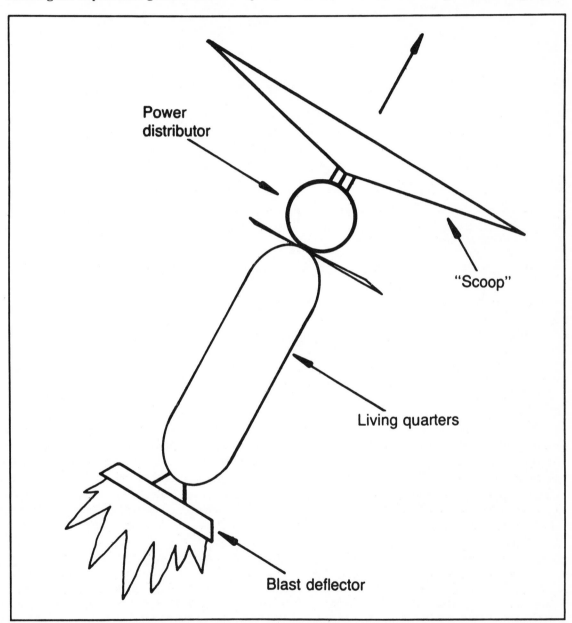

Fig. 5-15. The Bussard Ramjet may make interstellar, or even intergalactic, travel possible. The scoop picks up matter from space. The internal reactor ignites the matter and provides thrust. Proposed by R.W. Bussard.

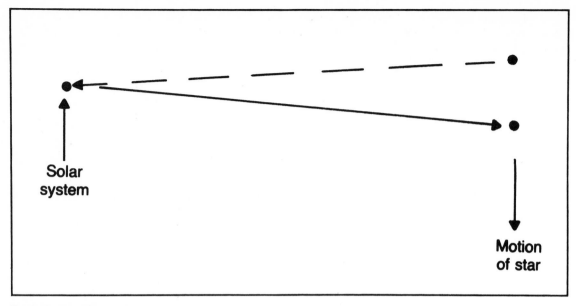

Fig. 5-16. A narrow-beam communications antenna would be hard to aim at a distant star. The star, and its planets, would move while the signal traveled across the gap.

from the antenna. The problem becomes like the proverbial search for the needle in a haystack. Nevertheless, we continue in the hope that the needle is there!

Some physicists and astronomers believe that there may be a mode of communication that we have not yet discovered, and that the truly advanced extraterrestrial civilizations are signaling via such esoteric means. Radio signals travel at the speed of light, but even in our immediate interstellar neighborhood, such transmissions require years to complete their circuit. We might send a radio message to the other side of the Milky Way galaxy: "Greetings from Earth!" A reply, if forthcoming, would not arrive for some 150,000 to 200,000 years.

Are there particles that travel faster than the speed of light? That is possible, and their existence has been predicted by some physicists. These particles are called tachyons. An advanced civilization might use tachyons to send messages in deep space. In fact, it has been suggested that the truly advanced societies would regard electromagnetic communications techniques as old-fashioned. There may be other modes of communication, too, such as four-dimensional electromagnetic waves,

for which the constraints of relativity do not apply. Even mental telepathy has been mentioned as a means for sending messages among the stars and galaxies!

Visitations

Some scientists think that intelligent civilizations would not simply spray out radio beams in the hope of being heard. It is possible that they might select certain stars and send automated space probes, at high speed, to orbit those stars and transmit beacon signals. This is the most elementary form of visitation. In the extreme, it is possible that beings from other worlds may someday land and present themselves to us. Perhaps this has already happened! Unfortunately, there is no indisputable evidence for such an occurrence in history.

Ever since written records were kept by mankind, there have been sightings of unidentified flying objects, or UFOs. Public interest in this phenomenon reached a peak in the twentieth century, when the dream of space travel became realizable, and interstellar journeys began to look like a real possibility.

Most UFO sightings have been explained in

terms of natural phenomena. The planet Venus, when it appears in the morning or evening sky through a thin haze, appears astonishingly large and bright, and has often been mistaken for a UFO. Airplanes, lightning discharges, and other common things have, at times, been regarded as UFOs. Some pranksters have manufactured false photographs of what looked like alien spacecraft. But a few UFO sightings have evaded all attempts at explanation.

If some of the unidentified flying objects actually are space ships piloted by extraterrestrial beings, we might be on the eve of contact with another civilization. But how long can we idly wait for them to come to us?

Some scientists believe that visits have already been made to our planet, but the details have been so twisted by legend, or erased by time, that we are not aware of what happened. Some religious writings, in particular, attract the attention of those who suspect such past encounters. The Book of Ezekiel, in the Old Testament of the Bible, tells of winged creatures and strange flying objects, which carried Ezekiel through the air.

Any civilization that has developed the technology to travel among the stars must have great patience. Perhaps such a society would conduct carefully planned expeditions to planets via space ships at regular intervals. Although the voyagers might not return for hundreds, thousands, or millions of generations, they would eventually come back with their stories. Records could be kept for an almost indefinite period of time. A stable civilization could keep in contact with other such societies, despite their separation in time as well as in space. We must be willing to accept even that which we, with our innate impatience, cannot understand. Some expeditions may have visited our planet at intervals of a few thousand years; if this is so, the most recent visit probably took place before the dawn of civilization, or before precise accounts could be made. Even a mere 1,000 years ago, we had very limited ability to make and preserve accurate historical records!

The search for extraterrestrial life will no doubt continue, and is one of the greatest pursuits ever undertaken by mankind. Eventually, if we try long enough and hard enough, we will probably make contact with any other civilizations that exist in our part of the galaxy.

MATTER AND ENERGY

Fundamental to our understanding of the universe, from its general structure to its smallest constituents, is the nature of matter and energy. Matter and energy are the media by which our minds communicate. An understanding of matter and energy will help us to build the apparatus we need to communicate through deep space, and ultimately to travel to other stars and galaxies. But our need for knowing matter and energy runs deeper than the purely physical: Matter is the substance of which our bodies are made, and energy is essential to our very consciousness. If we are to know what we are, and what forms life in the universe may take, we must know about matter and energy!

Albert Einstein, in his theory of relativity, was the first to specifically point out that matter and energy are manifestations of the very same thing. Matter can be converted into energy, and energy can be converted into matter. This happens quite often in particle interactions in our universe. But the questions "What is matter?" and "What is energy?" are not satisfactorily answered by simply stating that the two are different manifestations of the same thing!

As you hold this book, you realize (subconsciously, at least) that the book is comprised of literally billions of tiny particles, moving with extreme speed. What makes you believe this? Within the last 200 years or so, physicists have made calculations and carried out experiments that tell us that everything is made of these tiny particles. How do you see and comprehend the pages of this book? Tiny packets of energy, called photons, are emitted by a light bulb, or by the sun, and careen at 186,282 miles per second off the atoms that make up the page you are reading; some of these photons enter your eye, carrying the specific pattern of dark and light areas on the page. They pass through the lens of your eye, which is made of atoms; then, they interact with nerve endings to form electrical impulses. These impulses, themselves consisting of tiny, speeding particles called electrons, travel through complex groups of atoms to a huge congregation of other atoms—your brain. Then, somehow, you interpret the information contained in the book, hopefully in the way that I have intended. We call this reality! The more we know of reality, it seems, the less real it becomes.

The notion that matter is composed of tiny particles was put forth more than 2,000 years ago by the philosophers Democritus and Leucippus. They believed that matter might actually be made of indivisible constituents. This idea seemed consistent with the complicated nature of matter: its variable density, color, texture, and form. Only the manner in which the atoms are put together, thought Democritus and Leucippus, was variable; the particles themselves were all the same. Today, we realize that the "atoms" of Democritus and Leucippus are actually electrons, neutrons, and protons. These particles are indeed all the same. The passage of billions of years, or the location in the universe, has no effect on these particles. They have always been just as they are, and they act the same on the opposite side of the universe as they do on the earth. Had Democritus and Leucippus had access to a modern physics laboratory, they might have carried their theory further, but in their time, their minds and eyes were the only available instruments.

In the early part of the nineteenth century, scientist John Dalton again mentioned the idea of elementary constituents of matter. Then the experimental evidence began accumulating. The electron was actually discovered first, before the precise structure of atoms was known.

THE ATOM

An electric current, passing through a rarefied gas, creates a strange glow. You might conduct an experiment yourself to observe this, by means of a glass tube, a pair of electrodes, a vacuum pump, and a source of high direct-current voltage. Alternatively, you might simply conduct an experiment with a television picture screen.

When you bring a magnet close to the glowing, partially evacuated tube as it carries a current, the beam is bent. If you place a strong magnet near the face of a television picture tube, the picture becomes strangely distorted in the vicinity of the magnet. Even before the nature of atoms had been unraveled, scientists know that electric charges were influenced by magnetic fields. The vacuum-tube experiment was conducted in the late nineteenth century. Electric fields, as well as magnetic fields, were found to have an effect on the path of the current through the tube. It seemed that the glow in the tube was caused by individual charged particles, moving at high speed. The particles had a mass that could be calculated, as well as a definite speed that seemed to depend on the voltage in the tube. Because electricity seemed to be composed of these particles, they were given the name electrons.

The currents that flow in utility wires, making possible all of the conveniences of electric appliances, are made up of electrons. The image on a television screen is formed by electrons striking a specially treated phosphor. High-speed electrons continually arrive from space, bombarding our upper atmosphere. Every electron is the same as every other. They all carry exactly the same amount of electric charge, and they all have the same mass. In both of these respects, electrons are tiny objects! If you could gather up 5×10^{29}, or 500 billion billion billion, of these little charged particles, the whole collection would weigh 1 pound at sea level. In a 100-watt light bulb, approximately 5.4×10^{18}, or several billion billion, electrons pass through the filament every second.

It was not until the early twentieth century that atoms were found to consist of two particles, one of which orbits the other. The central particle, the nucleus, carries a positive charge that is a multiple of the charge on an electron, except opposite. Around the nucleus, electrons continually travel in orbit. The number of electrons is usually just enough to counterbalance the positive charge of the nucleus, so that the whole atom has a net charge of zero. The orbit of an electron around the nucleus of an atom is something like the orbit of a planet around a star. Rather than gravitational attraction,

170

as in the planetary system, electrical attraction between the unlike charges keeps the electron from flying away from the nucleus. Because of the resemblance of atoms to solar systems, some fanciful imaginings are possible: Are atoms actually solar systems, and electrons planets? Might some atoms be suitable for the formation of intelligent life? The idea sounds too silly to be true; it is also practically impossible to disprove!

The first, and simplest, orbiting-electron model of the atom was formulated by Ernest Rutherford. The most contemporary model of the atom is similar to his. The nucleus of the Rutherford atom is small and dense. The hydrogen atom is the simplest atom in the universe, with one unit negative charge in orbit around a unit positive charge (Fig. 5-17). The central nucleus is much heavier than the orbiting electron, just as the sun is much heavier than the earth.

More complex atoms exist, of course, and it was found that their nuclei always carry a positive charge that is a multiple of the charge on the hydrogen nucleus. However, the mass of the complex nucleus, while also a multiple of the mass of the hydrogen nucleus, is larger by a different factor. For example, the helium nucleus has two positive unit charges, but is about four times as heavy as a

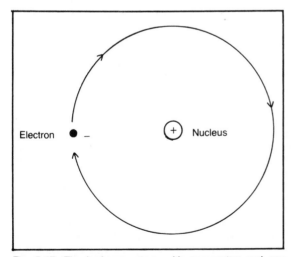

Fig. 5-17. The hydrogen atom, with one proton and one electron, is the simplest and most abundant atom in the universe.

hydrogen nucleus. Rutherford realized that this difference between the charge and mass factors could mean only one thing: Electrically neutral, as well as positive, particles must comprise the nuclei of the more complicated atoms. We call the positively charged particles protons, and the neutral particles neutrons.

The Rutherford model of the atom was refined by Niels Bohr of Denmark. Bohr theorized that the electrons in an atom always follow defined orbits, at specific distances from the nucleus. Today, we call these discrete orbits "shells." The shells are given names, represented by letters of the alphabet. The innermost shell is called the K shell; successively larger shells are designated by the letters L, M, N, O, P, and Q. Figure 5-18 shows this structural concept (not to scale). All of the shells are concentric; the nucleus lies at the center of each shell.

The electrons in every atom whiz around the nucleus with incomprehensible speed. If we were to call the period of revolution of one electron its "year," we would have to revise our time scale drastically in order to appreciate events in the atom. The electrons need not always stay in the same orbit, and in this way they behave much differently than the planets circling a star. An electron can gain energy and move into an orbit that is farther from the nucleus. An electron can lose energy and fall into a smaller, or lower, shell. But an electron cannot orbit between shells. Therefore, the energy gained or lost by an electron will tell us from which shell to which it has jumped. An infalling electron gives up a discrete packet of energy in the form of radiation. An outward-moving electron must absorb a packet of energy. The total possible number of shell jumps that any electron can make, in any particular atom, is finite. This behavior is the reason we see emission lines and absorption lines in the spectroscope as we observe a star, gas cloud, or galaxy in space. Energy is constantly being passed around among the atoms, resulting in the emission and absorption of energy at precise wavelengths. Because the spectrum of a distant star contains lines we can identify right here on our planet, we know that the atoms in the distant star are just like the ones we see in the laboratory.

171

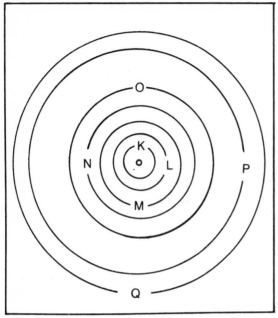

Fig. 5-18. Electrons may orbit an atomic nucleus only at discrete levels or "shells." The innermost level is called the K shell. Progressing outward, we find the L, M, N, O, P, and Q shells.

Now the connection between matter and energy begins to emerge! Electrons can emit or absorb energy packets, which we call photons. Photons can carry varying amounts of energy. The more energy contained in a single photon, the shorter the wavelength we observe, and the higher the electromagnetic frequency. Photons always travel at great speed along geodetic paths through the universe.

The greater the difference in the energy contained by an electron before and after a leap from one shell to another—a so-called quantum jump— the more energy is contained in the photon that is emitted or that must be absorbed (Fig. 5-19). Some energy transitions generate or absorb radio waves; some generate or absorb infrared, visible light, ultraviolet, or X-rays.

There are 92 elements commonly found in nature, ranging from hydrogen to uranium. Uranium normally contains 92 protons in its nucleus and 92 orbiting electrons. Still heavier elements can be produced under laboratory conditions,

but they are extremely unstable, and tend to break down into lighter elements. A cosmic limit is imposed on how large a nucleus can be. But elements can themselves be joined together; the nuclei of different atoms can share electrons, creating an entirely new substance, called a compound.

We are familiar with a number of different compounds. Water is the most recognizable compound on our planet; it is also one of the simplest. Two atoms of hydrogen combine with one atom of oxygen, and the orbiting electrons are shared among the three nuclei. Figure 5-20 is a diagram of the atomic representation of a water molecule. For water to be formed from hydrogen and oxygen, all we need is some hydrogen, some oxygen, and a high enough temperature. It is extremely fortunate for us that water is so easily created from these two elements. Life, as we know it, cannot exist without water!

There are many other familiar examples of compounds all around us. Iron oxide, a combination

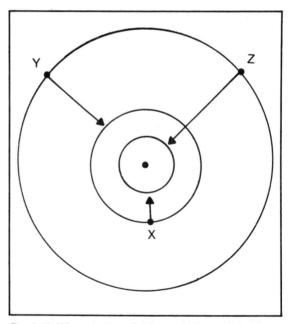

Fig. 5-19. When electrons fall from a higher orbit to a lower orbit, photons are produced at discrete wavelengths. If a photon of the right wavelength is absorbed by an electron, that electron will move into a higher orbit. These effects account for the emission and absorption lines in the spectra of distant objects.

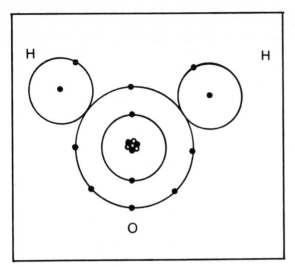

Fig. 5-20. Atoms of different elements can combine to share their electrons. Water is a common example of this: Two hydrogen atoms attach themselves to one oxygen atom.

of the elements iron and oxygen, is more often called rust. Copper oxide is a greenish compound that forms when the element copper is exposed to the air. Many compounds form in conjunction with the element oxygen. This is because oxygen readily combines with many other elements. Chlorine is like oxygen in this respect; it readily adheres to other atoms to form compounds.

Complicated compounds can be formed from the elements carbon, oxygen, hydrogen, and sometimes nitrogen. Some of these compounds are methane, ammonia, acetic acid, glycine, and alanine. More complex compounds include ribonucleic acid (RNA) and deoxyribonucleic acid (DNA). These compounds form huge molecules, but their fundamental constituents are the simple atoms. These large molecules can form most readily in the presence of amino acids and electric sparks, such as lightening discharges. It is believed that the conditions as on the "gas giant" planets—Jupiter, Saturn, Uranus, and Neptune—may be favorable for the development of amino acids and perhaps even RNA and DNA.

The most sophisticated molecules are capable of reproducing themselves, and this is a fundamental property of living things. Yet, even these molecules are composed of nothing more than electrons, neutrons, and protons in certain combinations.

ANTIMATTER

The idea that antimatter might exist is not especially new. But experimental evidence is much more convincing than purely theoretical considerations. The first antiparticle to be discovered was the positron, or positively charged electron. When high-energy photons strike the nuclei of atoms, debris often results. Among this debris, an electron is emitted, along with another particle, having the same mass as the electron, but opposite charge. This electron and positron are both created directly from the energy of the impinging photon.

If, however, the positron happens to run into an electron, both of the particles disappear, and a photon is created. The matter in the electron and positron are both converted directly into energy! It is as if the two particles cancel each other out in some way. For this reason, the positron is sometimes called an antielectron. If the electron is matter, then, the positron is antimatter.

In theory, if equal amounts of any kind of matter and antimatter are brought together, they will annihilate each other and produce energy. Following the discovery of the positron, a search began for other antiparticles. Eventually, antiprotons and antineutrons were also discovered.

Antimatter seems to have all of the properties of matter, in a mirror-image form. This has led some scientists to theorize that, for every material particle in the universe, there is one antiparticle. For every electron, there should be a positron; for every proton, there should be an antiproton; for every neutron, there should be an antineutron. If any particle meets with its counterpart, they will both be destroyed, and energy will result. This hypothesis leads to the question: Why, then, doesn't all of the matter in our cosmos meet with antimatter? When astronauts landed on the moon, they obviously found the same kind of matter of which they, and their space ship, were made; otherwise they would have been instantly annihilated, with a great burst of radiant energy.

In our part of the universe, at least, everything

is apparently made of matter, and not antimatter. We can be positive of this as far as the moon is concerned. We can also be certain in the cases of Venus and Mars, since we have landed unmanned craft on these planets. We can be reasonably confident that Mercury, Jupiter, Saturn, Uranus, Neptune, Pluto, and all of their satellites are composed of matter. We can be confident that the sun is composed of matter. But what about the distant stars and galaxies? The quantum jumps of positrons antimatter atoms would produce exactly the same photons as would the quantum jumps of electrons. The only way to be certain if an object is made of matter or antimatter is to touch it and see! If the universe contains equal amounts of matter and antimatter, then we should expect that some distant celestial objects are made of antimatter, rather than matter. But the idea that the universe must contain equal amounts of matter and antimatter is just that: an idea. Although the idea has a certain intuitive appeal, there is little other reason to believe it is correct, as far as we are presently able to tell.

In 1908, a tremendous blast occured over the plains of Siberia. The explosion caused trees to be felled outward for a radius of several miles, and the sound of the impact was heard for hundreds of miles. The actual "ground zero" impact site, however, contained no large crater or other obvious evidence of the impact of matter. If the object was a meteorite, it would presumably have left a sizable crater, but none was found. Many theories have been proposed in an attempt to explain this event. Some scientists think that a comet head exploded in midair. Some think that an alien spacecraft got out of control and destroyed itself to remove all traces of its presence. Some think that a piece of antimatter fell toward the earth, and was annihilated in the atmosphere. All of these are possible, but one thing is certain: The energy of this impact was almost beyond belief. A small chunk of antimatter could have produced it!

Is it possible that some of the galaxies we see are antimatter instead of matter? There is no way, from this distance, to be certain. How about the quasars? Could antimatter and matter be involved in mutual annihilation? This might account for their

tremendous energy and small size. For, when antimatter and matter meet, the conversion to energy is total. There is nothing left over. According to Einstein's famous equation, $E = mc^2$, the total conversion of matter to energy yields fantastic results; a matter-antimatter bomb would make the largest hydrogen weapon seem like a firecracker by comparison.

In our part of the universe, at least, matter has attained predominance over antimatter. While some particle interactions can create antimatter, such as the positron resulting from the impact of a photon against an atomic nucleus, the antimatter cannot last long until it is annihilated in a collision with a particle of matter. On the opposite side of the four-sphere universe, this condition might be reversed. But in terms of overall reality, there would be no difference. Their positive charge would be our negative; since these choices are arbitrary, the difference is not apparent until matter meets antimatter. If, or when, we travel to planets in other star systems, we might do well to conduct an investigation before we land. We may wish to send down an unmanned probe ahead of our shuttle. Then, in the event the planet is made of antimatter instead of matter, we will save ourselves a rude surprise!

PARTICLES WITHOUT END

Are electrons, protons, and neutrons the smallest particles? Evidently not. There are certain problems, even with the Bohr model of the atom. Complications arise that necessitate further detail in any model of the structure of matter. The quark model was invented to rectify these complications. The word "quark" comes from *Finnegans Wake* by James Joyce. Apparently, electrons are comprised of three quarks, and the same is true of protons and all other particles. Quarks can be categorized in three ways, which physicists call "up," "down," and "strange."

At this point, we ought to start wondering when this process will end, or if it will end! As you walk on a beach, you might look down and notice that the sand is composed of small particles or grains. Since you know a little particle physics, you realize that each of the sand grains is made of millions and millions of atoms. Most of the volume of these atoms is empty space. But the identity of the substance is determined by the unique combination of protons and neutrons in the nuclei of the constituent atoms, the orbiting electrons, and the manner in which the atoms are joined together as a compound. But on a still smaller scale, each electron, proton, and neutron is made up of quarks in special combinations of types we call up, down, and strange. What are the quarks made of? This brings us to a philosophical, more than physical, question. Could it be that every particle—no matter how small—is made of smaller constituents?

We can look at this situation inside-out: Is there a largest object? If our universe is shaped like a four-sphere, with finite surface volume, then we might say that the four-sphere is the largest object in our universe. But it is not especially difficult, once we have managed to envision the four-sphere, to imagine other such universes scattered throughout four-space. The idea that space is infinite is, perhaps, easier to deal with, intuitively, than the notion "This is all; there is no more." The same reasoning can be applied in the small-scale sense. It is easier to imagine that the particles of matter get smaller and smaller *ad infinitum*, than to think that there is a smallest particle! Scientists find it impossible to say, "This is all; there is no more."

One of the most far-reaching and fascinating theories in particle physics is put forth by Carl Sagan in his famous book, *Cosmos*. Suppose that a smallest particle does, in a sense, exist. That so-called elementary particle might be a complete universe, teeming with millions or billions of civilizations in various stages of evolution. They might have their own space-travel dreams, their own daytime and nighttime skies, perhaps of a different sort than ours. Our own universe, shaped like a four-sphere, could comprise a single elementary particle in a four-dimensional universe. Our entire cosmos might be too small to be seen, even with the most powerful microscopes in that larger universe! Of course, this is just a theory, but it cannot be disproved by us, with our scanty knowledge of re-ality. Such an idea illustrates how much farther we can progress. We are in no danger of reaching the end of cosmic reality.

A SHORT HISTORY OF EARTH

The evolution of intelligent life is a complex process, but not altogether outside our comprehension. We know that certain compounds, given the right circumstances, develop the ability to copy themselves. This process is the basis for life, according to the most contemporary scientific theories. Living things reproduce; nonliving things do not. A single particle, capable of splitting into two other particles indentical to itself, would rapidly spread on a planet of otherwise inanimate, nonliving particles. The animate-particle population would increase in geometric progression until the environment could no longer support it. Envision a single particle of this kind, reproducing once every day. At the end of the first day, there would be two particles. At the end of the second day, there would be four. By the end of the fifth day, there would be 32 of them. After two weeks, the animate-particle population would reach 16,384. After a month, there would be over a billion, or 1,000,000,000 of them. After twelve months, their number would be staggering: 2^{365}!

Let us now take an imaginary journey back billions of years in time, to the very beginning of life on our planet. Several billion years ago, the earth was very different than it is now. The atmosphere was a noxious mixture of chemicals we would find impossible to breathe: hydrogen, ammonia, methane, and water vapor. The oceans were less salty, and they were completely sterile. We would find some things strikingly familiar; the ocean waves would break on the continental shores with their characteristic, well-defined curl. The barren rocks would not look much different than some of our coastlines do today. But not a single tree would grace the horizon. No birds would soar over the water and the land. No grass would grow. Somehow, out of this environment, the earth developed, in and about three billion years, to a place of abundant life. No matter how it happened, it is hard not to say that it is a miracle.

According to modern science, life on earth began with complex groups of particles. Some molecules, called DNA and RNA, developed the ability to make copies of themselves. Not long after the first of these molecules appeared, the oceans of the earth were filled with them. It is believed that these molecules somehow stuck together, forming congregations of animate matter with more complex functions. But the exact way in which this happened is a mystery. The process has never been duplicated in the laboratory. The development of the first cells took billions of years, and we just do not have that much time to do experiments! There is some uncertainty, too, as to whether the cells actually evolved on this planet, or were transferred here from another planet. Perhaps, eventually, convincing details of the *modus operandi* of cell evolution will be found. We will have to find a way to compress millions of generations into a span of only a few years.

Two somewhat different kinds of cells appeared on the earth after about two billion years. One kind of cell was able to convert sunlight into the energy needed for its life processes; this cell was the first plant. Among the waste products of the plant cell was oxygen; other cells, specifically designed to use this waste for their own functions, developed. This cell was the first animal. The animal cell found, in the oxygen, a much more efficient source of energy than sunlight, for oxygen is a highly reactive element.

Some of the cells began to stick together in groups of two, four, five, or even a hundred. The reason why some cells clung to each other, while others did not, was probably accidental. A mutation in some cells resulted in their outer membranes getting sticky or rough. Large groups of cells were better able to survive, as it turned out, than single cells, and eventually the mutation became the norm. Single cells perished sooner than cells that were embedded in a clump. What would cause such a mutation? The behavior of even the tiniest particles is subject to a certain amount of random influence; occasional "errors" are made in any natural process. A high-speed helium nucleus, resulting from bombardment of the earth's atmosphere by

cosmic particles, might strike a cell and damage its DNA molecule. Or the DNA might reproduce "wrong." It is fortunate that such accidents happen. Otherwise, the earth would still harbor only the rudimentary beginnings of life. This is the principle of evolution and natural selection. We credit Charles Darwin with this theory.

The congregations of cells grew larger and larger. Some cells in a group were better equipped to perform certain tasks, for the benefit of the colony, than other cells. The outer cells, for example, were ideally suited to protect the inner cells against damage. The inner cells were better equipped to operate as food and energy processors. Natural selection thus dictated that the outer cells should be physically tough, while this was not required of the inner cells. Congregations with soft outer cells died, while those with hard outer cells survived longer and produced more offspring like themselves.

The process of natural selection requires much time. Traits develop to a refined form only after many generations. But the available time on our planet is billions of years; the sun, fortunately, is a relatively small star, and we expect that its life will be at least several billion years. Planets in orbit around giant stars, with much shorter lifetimes, would allow much less time for sophisticated life forms to develop, even if all the ingredients were there.

Precisely when, in the process of increasing complexity, did the collections of atoms and molecules cease to be simple matter? At what point can we call something "alive"? Some might say that the DNA molecule is alive because it reproduces itself. But others would say that the requirements for true life are more stringent. There is no well-defined point where we may say, "Now there is no life . . . now there is." Scientists would rather leave that decision to the theologians.

The more complex the groups of cells became, the more often mutations took place within a single organism. This is simply a matter of probability; mutations would occur twice as often in a group of two million cells as in a group of one million cells. For this reason, as life forms grew more varied and sophisticated, the process of evolution was accelerated.

Some of the cells from the primordial oceans were washed onto the shore, and no doubt the vast majority of them perished there. But some were able to survive in the puddles left by rain storms. These cells developed into the terrestrial plants. As the plants died, the soil was built up on the barren rock surfaces of the continents.

Some cells remained in the seas, where they developed into underwater plants and fishes of incredible variety. The oxygen-burning cell congregations developed the ability to propel themselves all about, in the never-ending search for food. From this point, the theory of evolution reached its climax. Some of the fish developed the ability to live on dry land. These animals evolved, over many millions of years, into the various giant lizards we call dinosaurs. The dinosaurs dominated the land, refining their bodies to almost complete perfection. The earth was much warmer, at that time, than it is today, and the cold-blooded lizards found survival easy. Some of the animals, much smaller than the dinosaurs, also survived, but with greater difficulty: They had to scurry about at night when the giant lizards could not see them. Their small bodies gave up heat quickly, and this problem was worsened by the cool nighttime temperatures. The result was the warm-blooded creature, or mammal.

About 35 to 50 million years ago, the climate of the earth began to cool. The reason for this is not known with certainty; there are several theories that have been formulated in an attempt to explain it. One possibility is that the activity of volcanoes, relatively dormant for millions of generations, increased, and the air was filled with dust particles that changed the heat-retaining characteristics of the atmosphere. Another theory suggests that the sun became cooler, and this idea is supported by the recent discovery of the dearth of neutrinos arriving from the sun (Chapter 2). Still another possible explanation is that the north pole, once located near the center of the Pacific Ocean, moved to its present location, resulting in dramatic changes in global weather patterns. Whatever the reason for the climatic change, the dinosaurs were unable to

deal with it. The event was too rapid for the process of evolution to function. The dinosaurs, no longer suited to the environment of their planet, perished within an astonishingly short time on the cosmic scale. By the year 10 million B.C., they were extinct. But the tiny, mouselike mammals survived.

There is no doubt that adversity plays an important role in the development of intelligent life. Without problems, there is no need for reasoning power. If the environment is perfect, evolution comes to a halt. The species reach a state of near flawlessness. This happened with the dinosaurs; after a certain amount of refinement in the stable climate of the earth, further mutations resulted in organisms that were less, not more, equipped to survive. The mutated creatures were always more likely to die. We may curse our bitter winters; but we owe our existence to them, according to the principles of evolution and natural selection.

The greatest of the ice ages began, and much of the temperate zone, previously a tropical jungle, turned to open savanna and arid steppe. Deserts formed. The level of the oceans dropped, exposing new land. The severity of the climate increased until glaciers formed, some as far south as the latitude of the central United States. The mammals, in this austere place, developed into more sophisticated and varied forms. It is believed that this process has culminated with the evolution of *Homo Sapiens*—man.

We have journeyed through billions of years. We have descended, according to modern scientific theory, from a single DNA-like strand of matter. What will the future bring?

THE NEXT FEW BILLION YEARS

Our species, *Homo Sapiens*, has roamed the earth for just a moment of cosmic time. We might get an idea of how quickly we have evolved and progressed by thinking of each billion years as a single day. On such a compressed scale, the big bang took place two or three weeks ago. Suppose it is now exactly 12 o'clock noon on the nineteenth day.

Our sun is less than a week old. The earth is about 4½ days old. The first life appeared only about yesterday. This morning, life had attained the status of the dinosaurs. A few minutes before noon, certain hairy creatures began to walk upright on two legs. The earliest civilizations fluorished within about two seconds of noon.

Christ was born during the zenith of the Roman Empire, about 0.4 second before noon. The long, dark Middle Ages lasted a miniscule 0.2 second, and ended a bit over 0.1 second ago. The two most terrible wars in history—those we call World Wars I and II—occurred within the blink of an eye. Now the clock strikes 12.

Technological progress has come quickly to us. Computers have enabled our species to effectively expand the brain by an enormous factor. We have unlocked the power of the atomic nucleus, illuminated our nights, and begun tentative ventures into space. Just 2/100 of a second ago on the above scale, someone was trying to figure out how to burn fossil fuels to make a horseless carriage.

It is clear that the evolutionary process has not changed our brains very much since the beginning of civilization. The natural process requires hundreds of thousands of years to modify an organism even to a small extent. In our present society, stone-age men and women are maneuvering sophisticated vehicles over the landscape, flying in computerized airplanes, and experimenting with powerful forces of many kinds. We tend to think that the technological revolution is here to stay, but it has been around for only a moment in cosmic time. What will take place in a million, or a billion, years?

Predictions for the future of *Homo Sapiens* can range from the most pessimistic to the most optimistic. The pessimists point to a theory called the Malthusian model of the evolution of a species. According to Thomas Malthus, a scientist of the early nineteenth century, any population that reproduces at a fast enough rate will ultimately face a crisis. Clearly, man falls into the category of a species that reproduces with extreme rapidity. Malthus showed that any population increase must take place in an exponential manner: the rate of growth gets faster and faster. If we plot the number of people in the world as a function of time, we get a graph that looks like the representation in Fig.

5-21. This graph might lead us to believe that the population can grow without limit; choose a number, no matter how absurdly huge, and there will, given enough time, be that many people! Of course, something must intervene, such as famine, plague, or war, to limit the population. Pessimists warn us that this point may be reached within a few decades. Optimistic proponents of the Malthusian hypothesis might give us a little more time—perhaps several centuries. But on a cosmic scale, that is hardly any time at all. Is Malthus really right? Or has he forgotten something?

On the optimistic side, we can point to the fact that the technological revolution has put off this kind of cataclysm many times already, and this might well continue. Effective and humane birth-control methods might be found. We are a species with reasoning power, and this gives us a great advantage compared to unthinking animals, such as the dinosaurs. We can expect the sun to keep shining for at several billion more years, before its hydrogen fuel begins to run out and our parent star swells into a red giant. If, within 100 years, we have come from horse-driven carriages to rocket-propelled spacecraft, it would surely seem that a few billion more years is enough time for us to find a new home in another star system! But clearly, our ability to survive depends directly on the strength of our capacity to reason.

The truth probably lies somewhere between the two extremes given above. We are not unthinking, automaton-like animals that behave purely by instinct; but intelligence has not yet evolved to total supremacy. The question "How long will we be a technically advanced people?" is well worth considering when we evaluate the probability that we will find another advanced civilization in the galaxy. If the pessimists are correct, and we have just a few decades left before we degenerate into oblivion, then we must presume that other evolving civilizations would face the same fate. It would be unlikely, then, that we would happen to detect such a society within the tiny moment of their spacefaring and communicating history. A ratio of 200 years to several billion years is not encouraging.

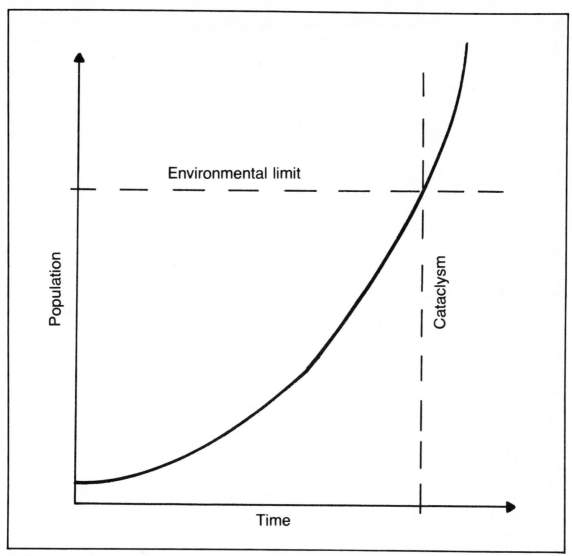

Fig. 5-21. The Malthusian model holds that a species must eventually face a cataclysm if the reproduction rate exceeds the death rate.

What are the real chances of locating another intelligent civilization in the universe? The answer depends, to some extent, on our own future, right here on this planet. Few would deny that we have just begun to tap the possibilities that physical science offers; given enough time, interstellar communication and travel could well become reality. If we can do it, so can other life forms. Conversely, if we find extraterrestrials wandering among the stars, our own confidence will receive a great boost.

THE GREEN BANK EQUATION

In 1961, a conference was held at Green Bank, West Virginia, the site of the Ozma experiments. The object of the meeting was to discuss and evaluate the possibility that there is other intelligent life in the Milky Way galaxy. The attending

179

astronomers, physicists, and other scientists developed a simple statistical equation, during the course of the conference, to define the number of advanced, communicating civilizations currently existing in our galaxy. This equation is usually known as the Green Bank equation. It has also been called the Drake equation, after the initiator of Project Ozma, and the Sagan-Drake equation, because of the participation of Carl Sagan (the now well-known author of *Cosmos*).

Several factors are involved in this mathematical model. Let R be the number of stars that form in our galaxy each year, on the average. Let P be the proportion of stars with planets of any kind; and let E be the average number of planets, per complete solar system, having conditions favorable for the development of life. Consider that L is the probability that life actually will evolve on a planet amenable to its development; let I be the probability that the life, once developed, will become intelligent; let C be the probability that this intelligent life will have the technology necessary to communicate through space; and finally, let us say that Y is the number of years that a capable society will try to contact other worlds. The probabilities, of course, are expressed as fractional numbers between 0 and 1. The entire product is called N—the number of civilizations presently undertaking to establish contact via radio. The complete equation thus looks like this:

$$N = RPELICY$$

We can immediately say one thing about the value of N: It is at least 1, since Project Ozma (and others like it) were conducted!

Star Formation: R

Our knowledge of galactic history is somewhat limited, especially when we look billions of years back in time. We might get a preliminary idea of the rate of star formation by considering the number of stars in our galaxy (about 100 billion) and the age of the Milky Way in years (about 10 billion). This would seem to indicate an average star birth rate of 10 per year. However, this is an oversimplification,

since the rate of star formation was probably different in the distant past than it is now. Our sun formed about 4 to 5 billion years ago, and it has taken that long for us to reach the communication stage. Perhaps the rate of star formation was greater when the sun was born.

If we are optimistic, we might choose a value of R as large as 20; the lower reasonable limit is probably about 2.

Planet Formation: P

The only positive example of a star that has orbiting planets is, of course, our own sun. Some evidence has accumulated to suggest that Barnard's Star, a close neighbor of the sun, has at least one planet. To get an idea of the likelihood that a given diffuse cloud of gas and dust will condense into a star with a planetary system, we need to look at the way such clouds behave. We examined the contemporary theories of stellar fomation in Chapter 2.

Almost all gas and dust clouds are believed to develop rotation, in much the same way as eddies form in a pool of water when a disturbance takes place. As a cloud contracts to form a star, the rate of rotation increases, and some of the angular momentum is transferred to regions well outside the star. This often results in the formation of binary or multiple star systems. Apparently, if the rate of rotation is significant, some of the angular momentum of a star must become concentrated in external bodies such as companion stars. But we in fact see many single stars, and these are good candidates for planetary systems. Our sun shed some of its angular momentum in a form that we now recognize as ten discrete elliptical regions. In eight of these ten zones, the gas and dust clumped together sufficiently to form planets. (Pluto might not have originated in the family of our sun; the asteroids apparently never had enough mass to aggregate and form a planet. Eight out of ten is not bad!)

About 10 to 20 percent of stars evolve singly; the rest become members of multiple systems. We might assume that almost all single stars have planetary systems, because of the tendency of contracting gas clouds to transfer some of their angular momentum outward. If this is true, we should adopt a value of P between 0.1 and 0.2. Some multiple systems might contain planets; some scientists have even suggested that P = 1!

Favorable Environment: E

In our planetary system, there is at least one planet suitable for the development of life, and that is the earth. There is some reason to believe that Mars might have rudimentary forms of life such as bacteria or viruses, but no verification has yet been made. We know very little about the deep atmospheres of Jupiter and Saturn; conditions at suitable altitudes on those planets might support the development of primitive life.

We naturally know much more about our own form of life, which is carbon-based, than we know about other possible bases of life. At low temperatures, silicon atoms can form long chain molecules similar to DNA, but it is believed that the lack of thermal agitation would result in an evolutionary process too slow to mature before the end of the universe. We have no idea whether there may be totally unexpected and unknown modes by which life might form. Therefore, to avoid over-optimism, we tend to take a somewhat narrow-minded viewpoint, and assume that life must resemble the only form we know.

If our solar system is typical, then we can expect to find an average of one earth-like planet per system—one hospitable abode for every five or ten stars! At the opposite extreme, we realize that the habitable zone around a star is a narrow region, and a planet could easily end up like Venus (at 800 degrees Fahrenheit) or Mars (far below zero), even if it were of ideal size. Some scientists think that the value of the factor E might be as large as 5. If we are still here when the sun begins to grow toward the red-giant phase, we will hope for a value this large! But it is more likely that E is a very small number, considerably less than 1.

To be suitable for life as we know it, a planet would have to be about the size of the earth, so that it could hold an atmosphere of the right density. It would have to be the correct distance from its sun. The tilt of the axis would have to be within certain limitations. There would have to be enough time for evolution to take place. Oxygen, hydrogen, carbon, and nitrogen would have to be present in sufficient amounts. But heavier gases, such as carbon dioxide, would have to exist only in small concentrations. This is a formidable set of constraints, and it is not likely that E can be much larger than about 0.5. It might be practically zero.

Development of Life: L

Even if a planet happens to present an ideal environment for the evolution of life, there is some chance that life will fail to be created. This term of the Green Bank equation seems at first to be inextricably linked with the E factor. If the environment is suitable for the formation of life, we might argue that this automatically sets L equal to 1. If all the substances are present, and all of the conditions are favorable, how could the value of L be anything but 1?

This view is a little too simplistic, for it overlooks the possibility of a cosmic disaster. Perhaps some freak accident might happen, such as an asteroid impact. A hospitable planet might be transformed into a planet unsuitable for the development of life. The composition of the atmosphere could change. The planet might be thrown into a highly elliptical orbit. The chances of such a catastrophe are uncertain, but even if we are pessimistic, it seems unlikely that it would happen to more than a few percent of planets. If we wish to be ultra-conservative, we might set the value of L at 0.5. The optimist would choose a value of nearly 1.

Development of Intelligence: I

The development of various traits, once life has formed in a particular place, is dictated by necessity. On the earth, life evolved to near perfection in the dinosaurs. These creatures were ideally suited to their habitat, and became the dominant form of life. They had little intelligence. The mammals eventually took over as the dinosaurs declined, and a primary factor in this event was the intelligence of the mammals. In one case, intelligence seems to have been unimportant for the evolution of a species to supremacy; in the other case, intelligence was essential. Which, if either, of these situations is the norm?

The answer to that question dictates the range of values we should realistically accept for factor I, the probability that intelligence will evolve on a planet once life has formed there. Some scientists believe, as we have seen in Chapter 2, that the sun fluctuates in overall energy output over periods of millions of years, and that this could have been responsible for the decline of the dinosaurs. If this is so, then we should expect that other stars might behave in a similar way, causing periodic climatic variations on their daughter planets. Then the conditions that led to intelligence on the earth were no accident, and we can believe that similar things would occur wherever there is life.

But how do we know that the solar-energy hypothesis of climatic change is correct? There may well have been some other reason for the cooling of the earth that took place millions of years ago. Volcanoes could have become more active, resulting in greater reflection of solar heat from the earth. This seems to have been the reason for an exceptionally cold summer that occurred in the year 1816. This phenomenon is not especially well documented, and few people know about it. During June and July of 1816, severe frosts occured in much of the northern United States, and snowfall even took place in some locations. The culprit is thought to have been a maverick volcano; dust from the volcano got into the upper atmosphere, and evidently this resulted in less solar energy reaching the ground.

Another hypothesis to explain the extinction of the dinosaurs is that the north pole was originally in the middle of the Pacific Ocean, and a cosmic disaster occured which caused the poles to change position. There is evidence of glaciers once having been in the region of Africa we now know as the Sahara Desert; this would have been the South Pole! Most of the United States, as well as Canada, Greenland, and South America, would have been near the equator if the poles were in Africa and the Pacific. Imagine the sun rising in what we now call the north, passing almost straight overhead, and setting in what we now call the south! What made the poles shift, if this theory is in fact correct?

Perhaps it was the drift of the continents, caused by forces from within the earth. Or it might have been the impact of a large asteroid. A deep trench in the Caribbean Sea is thought by some scientists to have been caused by the impact of an errant asteroid. Perhaps the asteroid was large enough to upset the orientation of the earth's axis.

Both the volcano and pole-shift events would have to be regarded as coincidental. If either of these explanations for the climate change on the earth is correct, we must accept that the change was a freak, an accident—and if it had not taken place, the dinosaurs would still reign supreme. We would probably not have evolved at all. Mammal development would probably have stopped at about the level of the mouse or squirrel.

The development of intelligence depends on the occurrence of adverse conditions. In this respect, intelligence is just like any other trait of a species, such as stereoscopic vision or flexible hands. But the degree of adversity must not be too severe. Otherwise, the environment of a planet could become too hostile to support any higher form of life. The value we adopt for the I factor depends on many things, and unfortunately, we do not have much data on which to make a good educated guess. If life develops in a certain place, the right amount of adversity will almost certainly result in the evolution of intelligence; what is the likelihood of this "ideal adversity" coming about? The pessimist would say it is almost zero. The optimist, having faith in the neutrino theory of the sun, would say the chances are high—at least 0.5.

Communication Technology: C

Even if a life form develops intelligence, it may not venture into space, either by means of electronic communication or by means of interstellar travel. We need only look at the dolphins, in our own oceans, to see how this kind of life might exist. While the dolphins are extremely intelligent—perhaps as intelligent as we are—they cannot build radio telescopes. They might not even be aware that other suns exist, simply because their eyes are equipped for underwater vision.

There are many possible reasons that an intelligent creature might develop, but not to the point of undertaking communications. Manual dexterity is certainly necessary if electronics or mechanical devices are to be constructed. We might find life forms much more intelligent than ourselves, but with absolutely no technological development. But what is the probability that a species will attempt communication with other worlds? It would seem to depend on the evolution of hands for manipulating such things as screwdrivers, wire cutters, soldering irons, and the like.

Conditions would have to dictate the evolution of hands. This, in turn, would necessitate that there be trees, rocky crags, or other complicated formations, in which such creatures could climb to escape their predators. But the creatures might run into caves instead. They would need keen eyesight, then, but they would not need hands. It is clear that any estimate of the value of C must be subjective.

In their book *Are We Alone?*, Robert T. Rood and James S. Trefil estimate the probability as 10 to 20 percent that an intelligent society will undertake communication. Thus we may say that C = 0.1 if we wish to be pessimistic, and C = 0.2 if we want to be optimistic. Trefil and Rood believe that some societies, even if they develop the ability to communicate by radio, may not desire to attempt interstellar contacts. This might be because their planet is covered by clouds all of the time, and the beings cannot see the stars. Radio and television could have developed on our planet, even if we had not known of the existence of other suns and other galaxies. It is also possible that a technological society might just not care to try to find life on other worlds. Even in our society, some people see as unjustifiable the expenditure of money and resources for such endeavors; there are more pressing issues.

Duration of the Attempt: Y

Let us imagine that somewhere, a technological society has evolved and has actually tried to make contact with another world. How long (in terms of earth years) will they keep trying if they do not succeed? Estimates can range from perhaps 10 years to however long we expect the society to exist. Clearly, the longer the period of time during which communication is attempted, the better the chances of success. A signal arriving at the earth 1,000 years ago would have fallen on deaf radio ears. The same would have been true just 100 years ago. It is frustrating to imagine that someone might have tried for 900 years to make contact with us, and then given up just a little too soon!

The pessimist might choose a value of Y = 10. The optimist might say that Y could be 1,000, 1,000,000 or even 1,000,000,000.

The Range of Solutions

We have made both pessimistic and optimistic guesses for the values of each factor in the Green Bank equation. Let us now proceed to determine the final product!

If we wish to take the extreme pessimistic point of view, we assign R = 2, P = 0.1, E = 0.001, L = 0.5, I = 0.001, C = 0.1, and Y = 10. This gives us the following number of societies presently attempting communication in our galaxy:

$$N = 2 \times 0.1 \times 0.001 \times 0.5 \times 0.001 \times 0.1 \times 10$$
$$= 0.0000001 = 10^{-7}$$

This is clearly an impossible value for N. We, the residents of the third planet in orbit around an ordinary star, have already tried to communicate with beings on other worlds. The value of N cannot be smaller than 1. One or more of the variables in the Green Bank equation must therefore be larger than the pessimistic extreme. Put together, the difference amounts to a factor of 10 million! Which of the Green Bank parameters are in error? What could be responsible for a miscalculation of seven orders of magnitude?

It could be that the values of E, I, and Y are much larger. These are the variables that allow for the greatest increases in the value of N. Both E and I could conceivably be 1,000 times as large as they are in the above product. The value of Y could be millions of times as great. It is possible to play with this equation for hours, finding many different combinations of parameters that result in N = 1!

Suppose, now, that we take the extreme optimistic viewpoint. Then R = 20, P = 0.2, E = 5, L = 1, I = 0.5, C = 0.2, and Y = 10^9. Then

$$N = 20 \times 0.2 \times 5 \times 1 \times 0.5 \times 0.2 \times 10^9$$
$$= 2,000,000,000 = 2 \times 10^9$$

This estimate is largely dependent on the value of Y. We have assumed, in this calculation, that a civilization could continue signaling and listening for a billion years. This is probably a gross overestimate. If there are that many civilizations trying to communicate throughout the galaxy, it seems that we should have heard at least one of them by now. But we haven't.

Who is Transmitting?

When we attempt to find an extraterrestrial civilization by radio, our first impulse is that we are much better off listening than sending. It is not the nature of the scientists to take "shots in the dark." Blind transmissions, it seems, would serve little purpose. For this reason, virtually all of our time, in the search for communicating interstellar societies, has been spent scanning the heavens with radio receivers. Few transmissions have been made. A reply, if any, to a transmission would not come for years. In many cases, the reply would not come for generations or even centuries! But are transmission really a waste of time? Perhaps this reasoning is too simplistic. Consider the following analogy.

Amateur radio operators are familiar with a strange phenomenon called the "dead-band syndrome." The radio amateur can sit at his operating desk at any hour of the day or night, and talk to other radio amateurs using a transmitter and a receiver. There are several different bands of frequencies from which to choose. Depending on many factors, long-range communications may or may not be possible on a certain band at a given time. But these conditions can change quickly.

Imagine a scenario in which a radio amateur sets his receiver to a certain frequency band and hears nothing. This could be because conditions are not good for communication at that frequency at that time. But there is another possible reason for the absence of signals. Perhaps 100 other radio amateurs are sitting at their station consoles, listening on the same band of frequencies. They all come to the conclusion that the band will not support communication. But how do they know that if nobody makes a transmission?

One person dares to waste one minute and send "CQ" (meaning, "Calling anybody!"). His call is heard by three other stations. The "dead band" comes alive! The other operators tune their receiver dials, and hear the conversation. Then they, too, realize that conditions are favorable. More stations send CQ, and more conversations arise. The initial hunch was wrong! Everyone thought that conditions were poor for communication, so no one tried to make contacts—until one person dared to question his intuition. Had he not done so, there might have been no conversations at all.

Let us apply this reasoning to our interstellar communications. Suppose that there are thousands, millions, or even billions of civilizations listening for signals from other worlds. Who will transmit first?

Dr. Frank Drake, the initiator and conductor of Project Ozma, has already come to the conclusion that someone must do the transmitting, and that it might as well be us! In 1974, Drake used the massive radio telescope at Arecibo, Puerto Rico, to transmit a simple message. This transmission has already passed some of the nearer stars. The message that Drake sent was easily deciphered by several of his friends. The code that Drake used is a form of binary transmission, regarded as one of the simplest—and most efficient—ways of sending an electromagnetic message. Our civilization has sent a celestial CQ! The reply, if any, may not come for centuries; but no one would hear our call if it were never made. We cannot succeed in contacting other worlds unless we make an effort.

The signal that Drake sent consists of 73 individual rows of information. Each row contains 23 bits, or binary digits. The numbers 73 and 23 were chosen partly because they are prime numbers: They cannot be broken down into multiplicative constituents. Other prime numbers could just as well have been chosen; these two seemed especially convenient. The use of prime-number quan-

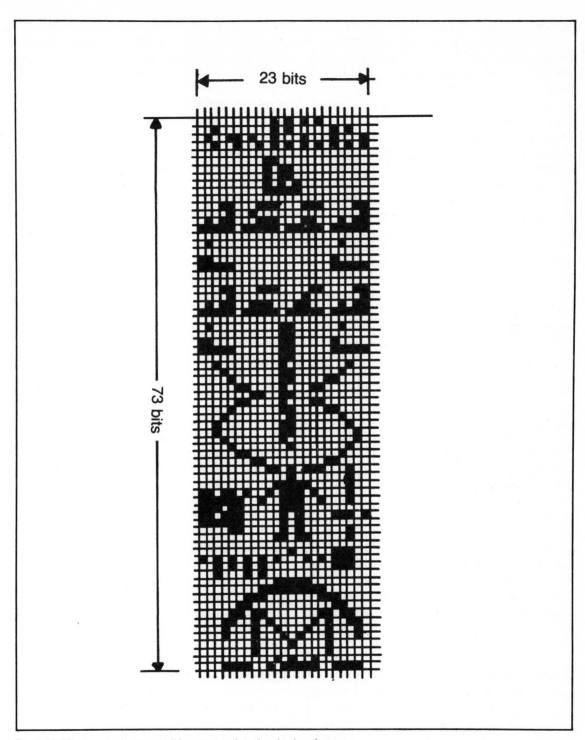

Fig. 5-22. We humans have sent this message into the depths of space.

tities for the rows and columns minimizes the difficulty of decoding the message. An intelligent radio-receiving operator would record the message row by row, until the graphical pattern looked like the drawing of Fig. 5-22.

The top of Drake's visual message contains a numerical count from 1 to 10 in a special binary form. The next line contains the atomic numbers of the essential constituents of earthly life. These are hydrogen, with atomic number 1, carbon, with atomic number 6, nitrogen, having an atomic number of 7, oxygen, with an atomic number of 8, and phosphorus, with an atomic number of 15. The next line contains a binary representation of the chemical formula for deoxyribonucleic acid, or DNA, the fundamental component from which animal life has arisen on our planet.

Below the DNA representation, a stick figure is shown, having the shape of a human body. Other binary digits give the number of people on the earth, the intensity of the signals used to transmit the message, and a pictorial diagram of the solar system showing all nine planets and their approximate sizes. The stick figure is positioned exactly above the third planet. At the bottom of the message is a representation of the Arecibo radio telescope itself, and information about its physical size.

The search for signals from space, meanwhile, continues as funds are made available. In 1977, Dr. Bob Dixon and his colleagues, of Ohio State University, noticed what appeared to be an artificial signal in their radio-telescope computer printout. The nature of this signal has never been resolved. Unfortunately, when the same part of the sky—the constellation Sagittarius, the archer—was scanned again, the strange signal was no longer there. But the printout still bears the handwritten remark in the margin next to the unidentified signal: "Wow!"

SPACE LANGUAGE

For a signal to propagate over distances of hundreds, thousands, or millions of light years, it must either be immensely powerful, or its energy must be highly concentrated. The generation of the extreme power required to send a haphazard, wide-band signal to other star systems is past the capability of mankind. Literally trillions of watts would be needed. The signals we send must therefore be directed into a narrow beam, and must be confined to a tiny band of frequencies. This, in turn, forces us to use the most efficient possible language when we send a celestial CQ.

It is possible that extraterrestrials might develop radio telescopes so sensitive, and with such resolving power, that they could intercept our radio and television broadcasts. A radio telescope of this calibre would probably require an interferometer consisting of antennas on several planets within a solar system! The probability that this highly directional system would be aimed at our planet, and that the receiver would be turned to the correct frequencies, is miniscule. Most experts in celestial communications seem to agree that our signal must be deliberately tailored for interstellar propagation.

The binary system is the most efficient method of modulating a radio signal. A binary signal is either all the way on, or all the way off. There are many different binary codes. The International Morse Code, familiar to radiotelegraphers and amateur radio operators, is a form of binary communications medium. Figure 5-23 illustrates the binary representations of the letters of the alphabet, the numerals 0 through 9, and some punctuation marks in the International Morse Code. The "dot" consists of a single bit, and the "dash" consists of three bits in a row. The space between letters is three bits; the space between words and sentences is seven bits. The International Morse code is not a very efficient binary system, however. The longest characters have 19 bits, not including the following space. This is much longer than necessary. Another problem with the Morse code is that the characters have variable length. The letter E, for example, is one bit long, but the numeral 0 has 19 bits. This would probably confuse an extraterrestrial. It would be rather difficult to decode a message sent in Morse.

A character length of just six bits allows us to construct up to 2^6, or 64, different symbols in a binary language. This gives us room for the 26 letters of the English alphabet (or the larger

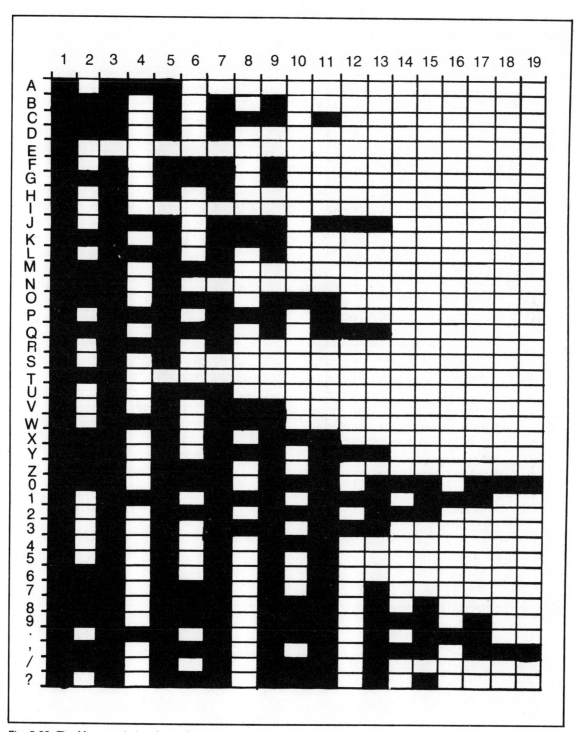

Fig. 5-23. The Morse code is a form of binary code with variable character length.

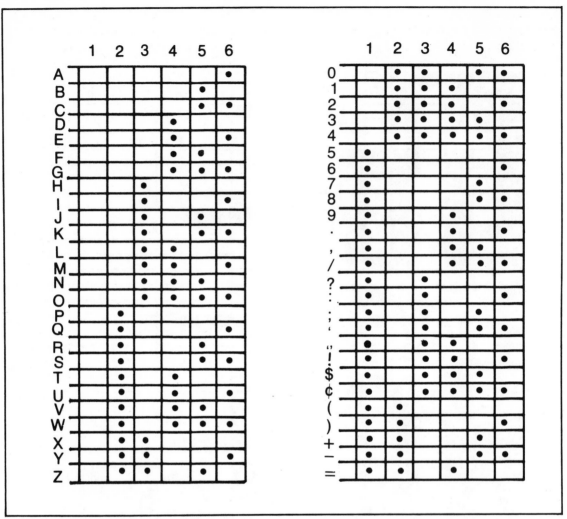

Fig. 5-24. A binary digital code with constant character length, useful for sending interstellar or intergalactic messages.

number of letters in some other alphabets); we have room for the ten numerical digits 0 through 9; there is plenty of space for all of the punctuation marks. The remaining binary combinations might represent mathematical symbols or special abbreviations. Figure 5-24 shows one possible six-bit binary communications code that we might use for the transmission of messages through space. This code would result in more efficient use of the transmitter power than the Morse code; we might send more information in the same amount of time. Alternatively, we might reduce the number of

transmitted bits per second, improving the signal-to-noise ratio and extending the range of transmission. The code depicted in Fig. 5-24 has the advantage that all of the characters are the same length. This maximizes the ease with which a message can be deciphered.

Of course, there is nothing special about the particular choice of character representations in Fig. 5-24. However, we know that only six bits are needed to uniquely determine a character in the English alphanumeric system. If we are willing to spell out numbers, rather than writing them as numerals, we can get away with just five bits per character: There are 32 different combinations in a five-bit binary code.

When we send a cosmic CQ, we will always use binary codes. We expect that, if we receive a message from an extraterrestrial civilization, that signal, too, will be in binary form.

WE'VE BEEN WAITING

Suppose we take the most optimistic possible viewpoint for the future of our race, and hypothesize that we will dominate the earth until the sun grows old. The sun will remain well-tempered for a few billion more years; it is a healthy, middle-aged parent star. We have plenty of time to get ready for the journey to another star system. But some day, we will have to make that trip.

Today, we look at the cosmos as a fascinating place to visit. People have longed to travel to the stars. But there is no pressing need for it now. Desire is not the same thing as necessity. A few billion years from now, our descendants will gaze at the stars with a different attitude: They will be looking for a new home.

This turning point must confront any civilization that survives long enough. Some stars have already died. If those stars had planetary systems with enduring, intelligent life, then there are prob-ably space travelers somewhere! The instinct for survival would drive any advanced race, faced with doom on their own planet, into space.

Where would they go? If their physiology is anything like ours, they might come to the earth. But perhaps there are other suitable planets that are not yet populated by intelligent beings. For a race to endure for billions of years, the process of evolution would have to cultivate gentleness; war-making would become contrary to survival, where once it had been a favorable trait. The fittest would become the least fit. Can natural selection make such a reversal? Our planet is already occupied. The extraterrestrials would probably rather leave us alone. They might even fear us!

Is the species *Homo Sapiens* by itself in the galaxy or in the universe? We keep asking ourselves that question, and we keep searching for some reassurance that we are not alone. Perhaps, however, the real answer lies within ourselves. If we are to become star travelers, we will have to cooperate more closely than in the past. The technological feat of interstellar travel can be accomplished only by a race that works together. It would be reassuring to find another civilization that has endured and reached for the stars.

When the first starship from Earth accelerates past the orbit of Pluto, perhaps its crew will receive the message: "We've been waiting."

Suggested Additional Reading

Baker, Robert H. *An Introduction to Astronomy.* Princeton, NJ: D. Van Nostrand Co., Inc., 1961.

Born, Max. *Einstein's Theory of Relativity.* New York: Dover Publications, Inc., 1965.

Caidin, Martin. *Destination Mars.* Garden City, NY: Doubleday & Co., Inc., 1972.

Gibilisco, Stan. *Understanding Einstein's Theories of Relativity: Man's New Perspective on the Cosmos.* Blue Ridge Summit, PA: Tab Books Inc., 1983.

Hey, J.S. *The Evolution of Radio Astronomy.* New York: Science History Publications, 1973.

Jastrow, Robert. *God and the Astronomers.* New York: W. W. Norton & Co., Inc., 1978.

Jastrow, Robert. *Red Giants and White Dwarfs.* New York: Harper and Row, Publishers, Inc., 1967.

Jastrow, Robert. *Until the Sun Dies.* New York W. W Norton & Co., Inc., 1977.

Ronan, Colin. *Invisible Astronomy.* New York: J. B. Lippincott Co., 1969.

Rood, Robert T. & Trefil, James S. *Are We Alone?* New York: Charles Scribner's Sons, 1981.

Sagan, Carl. *Cosmos.* New York: Random House, 1980.

Sullivan, Walter. *We are Not Alone.* New York: McGraw-Hill Book Company, 1964.

Taylor, John G. *Black Holes.* New York: Avon Books, 1973.

Index

Index